PRAISE FOR THE BOOK

There is no one who writes quite like Andrew Morris. This is a great collection of chapters on everyday science. If only more of us had been taught like this at school, our universities would be bursting with people trying to get in to study science.

—**Michael J. Reiss**, Professor of Science Education, University College London, UK

I wish science in school could have been this fascinating and accessible. This book is full of 'mind blown' moments! I love that it's about learning about the world through curiosity and discovery, not fusty textbooks. I love learning about the science behind everyday things without it being too complicated – or patronising. Perfect for dipping into on the train in the morning!

—**Hilary Davies**, Civil Servant, Kent, UK

With this book, Dr Morris has opened doors for the everyday person who may have thought that science was only for the brainiacs of this world. His desire to bring us all along, coupled with his cheery down-to-earthness, is refreshing and welcoming. A gem of a book!

—**Janet Johnston**, curious adult, Florida, USA

In this book, Dr Andrew Morris admirably achieves what he sets out to achieve – making science accessible to readers like myself, who know very little about any one of the sciences, and to bring them alive in just the way that suits the interested amateur, starting with questions we are likely to ask, proceeding by way of vivid example, plain English and exactly the right amount of science to take us all the way with him to gain the understanding we were looking for. The pitch is perfect: nothing patronising and no dumbing down, but nor are there demands on readers that only natural-born scientists are likely to meet. Dr Morris is a lucid and persuasive advocate, with many years of experience of working with groups of adults doing just what he is doing in this book – helping adults to understand science for themselves. I cannot imagine a better introduction to bugs, drugs and three-pin plugs than what is contained in these pages.

—**John Vorhaus**, Professor of Moral and Educational Philosophy, University College London, UK

Mission accomplished! A passionate teacher determined to build confidence and knowledge in those whose past encounters with science were a turn-off, Andrew succeeds in both with his engaging and accessible book. It certainly worked for me!

—**Liz Walton**, retired Principal of William Morris Sixth Form, London, UK

Andrew Morris has that rare talent for using everyday observations and experiences to unpick and explain the most complex of scientific ideas. He does so in this book, as in previous works, by tapping into our natural curiosity about the world around us. An explanation of why our ears pop during a flight and how a crisp and silvery scene was captured in the pond of a winter garden lead us to understand the science behind solids, liquids and gases.

—**Ian Nash**, freelance journalist, senior partner in Nash&Jones Partnership and former assistant editor of the TES, UK

Bugs, Drugs and Three-pin Plugs

For the millions who remain curious about the world around them, but gained little from science at school, this book offers a way forward. Based on live discussions with adults from all walks of life, each chapter begins with an everyday experience, like swallowing a pill or watching a bee on a flower. The main scientific ideas underlying each topic are then explored, so that understanding of a set of fundamental concepts builds up gradually throughout the book.

In contrast to more traditional approaches to science learning, topics range freely across the subject areas. The story of Covid, for example, includes aspects of biology, chemistry, mathematics and social behaviour. Plain English is used throughout, and mathematical expressions are avoided. Key points are illustrated with clear diagrams and photographs.

By drawing on questions and perspectives of ordinary people, the book offers an introduction to basic ideas in science as a whole, rather than any one particular subject. For the adult wishing to make good a gap in their understanding, it provides a starting point for entering the rich world of popular science.

ANDREW MORRIS

Bugs, Drugs and Three-pin Plugs
Everyday Science, Simply Explained

CRC Press
Taylor & Francis Group
Boca Raton London New York

CRC Press is an imprint of the
Taylor & Francis Group, an **informa** business

First edition published 2023
by CRC Press
6000 Broken Sound Parkway NW, Suite 300, Boca Raton, FL 33487-2742

and by CRC Press
4 Park Square, Milton Park, Abingdon, Oxon, OX14 4RN

CRC Press is an imprint of Taylor & Francis Group, LLC

ISBN: 978-1-032-22494-7 (hbk)
ISBN: 978-1-032-22492-3 (pbk)
ISBN: 978-1-003-27277-9 (ebk)

DOI: 10.1201/9781003272779

Typeset in Joanna
by codeMantra

For those who never stopped asking questions.

Contents

Acknowledgements   ix
Author   xi

Introduction   1

How Do Pills Know Where to Go?: *The Shape and Nature of Drug Molecules Dictate Where They Act*   **One**   3

Singing and Navigating – The Extraordinary Feats of Ordinary Birds: *What Science Tells Us about Bird Song and Migration*   **Two**   9

The Colour of Light: *The Nature of Light and Colours of the Spectrum*   **Three**   19

Why We Look Like Our Parents – A Bit: *The Story of Genes, Chromosomes and DNA*   **Four**   29

Atmospheric Rivers: *Rain, Humidity and the Water Cycle*   **Five**   41

The Perennial Question of Sex: *How Plants Reproduce*   **Six**   51

The Geometry of Ice: *What Lines on an Icy Pond Reveal about the Underlying Structure of Matter*   **Seven**   61

Very Small and Very Busy: *Life Inside the Cell What Goes on Inside Our Human Cells*   **Eight**   71

Enzymes: *Familiar to Brewers and Cheesemakers, but What Are They?* **Nine**   83

Sugar, Carbs and Type 2 Diabetes: *The Substances behind the Surge in Diabetes 2* **Ten**   93

Sticking Together: *The Science of Adhesion* **Eleven**   101

A Nice Warm Shower: *What's the Difference between Heat and Temperature?* **Twelve**   115

Why Your Ears 'Pop': *What Popping Tells Us about Pressure, the Middle Ear and the Atmosphere* **Thirteen**   123

Making Decisions: *Evidence from Experimental Psychology about How We Make Choices* **Fourteen**   129

COVID-19: Viruses, Lungs and Epidemics: *The Basic Science of the Pandemic* **Fifteen**   139

COVID 19: Immunity, Vaccines and Variants: *The Immune System and How Vaccines Stimulate It* **Sixteen**   151

Energy and the Climate Emergency: *Greenhouse Gases, Energy and Heat Pumps* **Seventeen**   165

Electricity: *What It Is and Where It Comes From* **Eighteen**   179

Reflections **Nineteen**   193

Acknowledgements

The starting point for each chapter has been a question posed, or observation made, in a group discussion or conversation with friends or colleagues. I am indebted to all who have contributed in this way: Peggy Aylett, Jane Brehony, Hilary Davis, Kristina Handy, Jan and Bob Johnston, Debbie Karp, Carmen Kearney, Susan Kearney, Will Lake, Victoria Minton, Paul Morris, Alison Morris, Kate Oxley, David Oxley, David Rowley, Melissa Rosenbaum, Linda Slack, Liz Walton, Emily White and Anna Wodjowicz.

I am very grateful to the teachers and researchers who have given specialist advice on particular chapters, including Peter Campbell, Dr Reha Celikel, Dr Jay Derrick, Professor Kathleen M. Galotti, Professor Nicholas Lesica, Dr Ralph Levinson, Dr Kieran Quill, Professor Michael Reiss, Geoff Stanton, David Swinscoe, Professor Andrew Tolmie and Professor John Vorhaus.

The editorial team at CRC has been highly supportive and responsive throughout the production process. I am extremely grateful to Alice Oven and Shikha Garg for guiding me through it.

Finally, I thank my partner Franco Carta for his patience and forbearance during the writing of the book.

Andrew Morris was a teacher of physics and mathematics for many years in Further Education and Sixth Form Colleges. He subsequently set up an experimental course for adults with little background in science at the Mary Ward adult education centre in Bloomsbury, London. Sessions started from the questions participants asked and followed the course of discussion they chose, rather than adhering to a fixed syllabus. He has continued to run these discussion sessions informally for over 20 years (online during the pandemic). Records of these discussions form the basis of this book.

Dr Morris also works on the use of pedagogical research to inform teaching. In this competancy, he became a research manager at the Further Education Development Agency and Director of the National Education Research Forum under Sir Michael Peckham. Dr Morris has a degree in physics from UCL and a PhD in molecular biophysics from the University of Leeds. He is an Honorary Associate Professor at UCL Institute of Education and a former President of the Education Section of the British Science Association.

Science doesn't have to be hard; or at least, not harder than other subjects. Sadly this is not how it seems for many people. For too many, previous encounters with science have been bruising. Studies reveal widespread negative feelings about science at school. Positive attitudes to science in general, however, are found in surveys of adults. My own experience with informal discussion groups for adults with little knowledge of science bears this out. Curiosity about the world around us seems to remain strong for many, whatever their background. The proliferation of books and TV and radio programmes on popular science are testament to this. It is this unquenchable spirit of enquiry that motivates the writing of this book.

The purpose of the book is to help the generally literate person grasp some fundamental ideas in science, much as they may have done for politics, history or sociology, for example. The aim is to offer an imaginative sense of what is meant by a molecule, a cell, a virus, energy, temperature or magnetism without invoking formulae or excessive factual detail. To do so, it draws on hundreds of live discussions held over several decades with adults keen to develop their understanding, despite having a weak background in science. The approach differs so much from traditional learning in science that it almost turns it upside down. In place of a fixed syllabus of topics in biology, chemistry and physics, discussions start from questions people actually ask about the world.

The chapters of this book reflect this reversal. They begin with questions about icy ponds, superglue or vaccination, then move on to the underlying science of the behaviour of water molecules, nature of adhesion or workings of the immune system, for example. Natural curiosity does not confine itself within the boundaries of school subjects. Questions about diet take us into organic chemistry, cell biology and nutrition; observation of migrating birds leads into anatomy, magnetism, astronomy and ornithology. Furthermore, the chapters of the book, like the discussions upon which they are based, tackle a great variety of

DOI: 10.1201/9781003272779-1

topics. Each chapter ranges across aspects of the topic as they crop up in everyday life, rather than in a text book. For the person seeking detailed specialist knowledge on a topic or a logically linked set of topics, this book is to be avoided. Plenty of other books serve that purpose well. A science specialist hoping to see their subject laid out fully and thoroughly will be equally disappointed. Instead this book touches on an unusually wide range of subjects in order to introduce a good number of the fundamental concepts that underlie the questions people ask. It aims to help the reader develop some idea of the realm of science, not just one particular branch, such as genetics or meteorology.

Science can appear to be hard for a variety of reasons. Terminology may be unfamiliar and explanations poorly communicated. As with any fascinating and rich subject matter, it's all too easy for books or teachers to present too many new ideas, too rapidly. The neuroscience of learning tells us that our working memory can only deal with a limited amount of new information at one time: overloading it can be as defeating as any inherent difficulty in the material. With these points in mind, most chapters have been kept relatively short and technical terminology avoided wherever possible. It is hoped this approach makes the material more accessible, though it does mean that some fascinating aspects of a topic may not be covered. You may well end a chapter full of further questions. The purpose of the book is to introduce scientific ideas rather than to elaborate them. If it succeeds in this it may give you, the reader, confidence to explore topics of particular interest in greater detail in one of the many good books on the popular science shelves.

One

How on earth do pills know where to go? A question that many of us will have pondered at one time or another. What happens to an aspirin once it's been swallowed? If it's intended to cure your headache or regulate your heartbeat, how does it know where to go? Is it tagged in some way? Does it carry some kind of destination board, like a bus: 'Frontal Cortex' or 'Coronary Arteries'? Of course not – how could it navigate the route even if it did?

The answer to this question has only begun to become clear in recent decades, as the nature of the molecules in the body and the medicine cabinet has become understood in some detail. It was their precise three-dimensional structures that provided the key. With the aid of X-ray crystallography in the mid-twentieth century, it was established that molecules, whether of medicines or bodily tissues (or almost any other kind of substance), had tightly defined structures, with all the atoms of which they were composed spaced apart at precise distances from one another, oriented at precise angles.

Figure 1.1 is a diagram of a typical small molecule – caffeine. It shows one way in which the structure of a molecule can be represented. A molecule is made up of a group of atoms closely linked (bonded) to one another. Molecules found in living systems contain many carbon atoms – so many that they aren't always indicated, they are just implied where lines meet at a vertex.

The labelled atoms are H – hydrogen, C – carbon, N – nitrogen and O – oxygen. The length of the lines and the angles between them correspond to the exact distances and angles between atoms, as revealed by X-ray crystallography and other techniques.

Each type of molecule has its own particular shape, and this helps determine the properties of the substance of which they are part. Two examples, out of countless possibilities, illustrate the point. Rubber, as we know, is able to twist and flex. This is a direct consequence of the long chain-like shape of the molecules of which it is made. Normally tangled

DOI: 10.1201/9781003272779-2

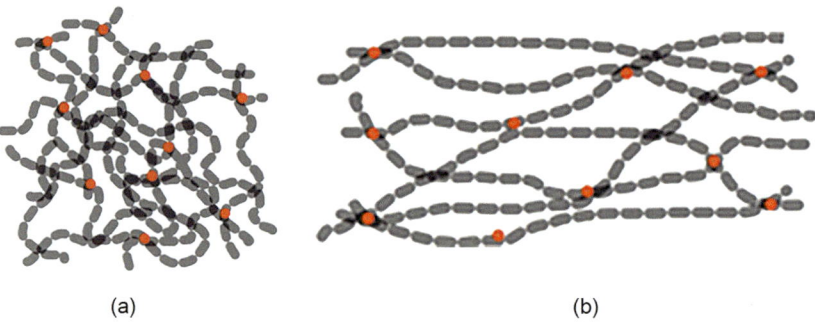

Figure 1.1 Diagram of a caffeine molecule.
(Image credit: Wikimedia, https://creativecommons.org/licenses/by-sa/3.0/deed.en)

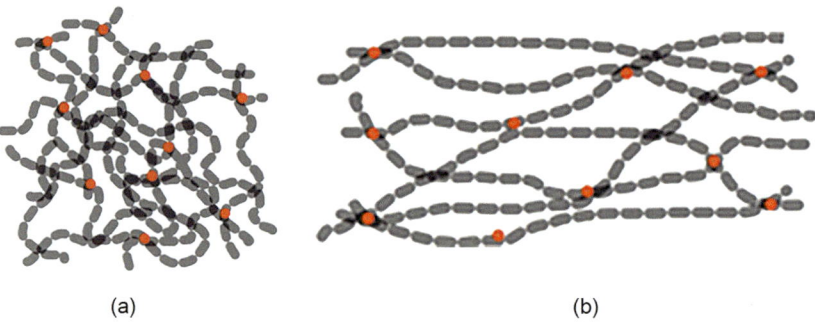

(a)   (b)

Figure 1.2 Diagram of rubber molecules. (a) relaxed (b) stretched.
(Image courtesy of The Naked Scientists.)

together, as illustrated in Figure 1.2a, they can be disentangled, forming a more orderly linear   arrangement, when the rubber is stretched (Figure 1.2b).

A very different molecule called retinal, located in the retina of the eye, enables us (and other species) to see. The usual shape of the molecule is as illustrated in the upper part of Figure 1.3 labelled (a). When light falls on it, however, the molecule's shape changes, quite dramatically (Figure 1.3 lower part, b). The symbol γ (gamma) represents the light energy that is causing the transformation.

On close examination, you'll see that the two molecules A and B are identical on their left-hand sides; on the right, they only differ in that one bond has twisted round, placing the right-hand end in a different position in B compared to A. This simple change – the rotation around just one bond between a pair of atoms – changes the shape of the molecule profoundly. It is this simple alteration in the structure of the molecule that leads to a signal being sent along the optic nerve to the brain, giving us the sensation of vision.

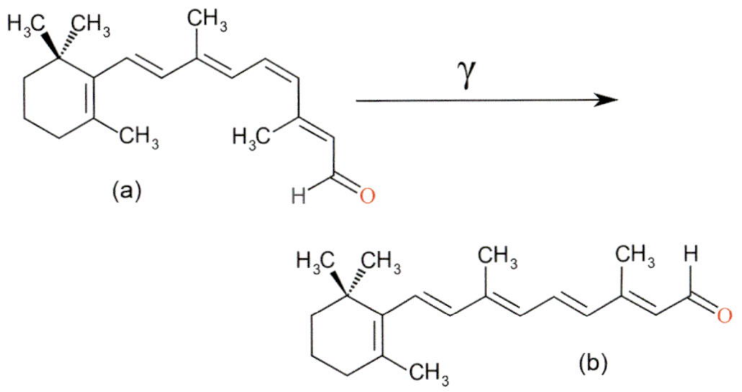

Figure 1.3 Diagrams of the retinal molecule before (a) and after (b) light falls on it.
(Image credit: RicHard-59 via Wikimedia, https://creativecommons.org/licenses/by-sa/3.0/deed.fr)

For medicines, the crucial question is what happens when two molecules interact with one another; how does a molecule of, say, aspirin interact with the appropriate molecules in your body? This is where shape matters. The molecules of many medicines are relatively small, comprising a few dozen atoms each. Figure 1.4 shows models of three well-known ones. This type of model represents the relative size of each atom (carbon is grey or black, oxygen is red, hydrogen is white and nitrogen is blue).

Many of the molecules that make up body tissues are, by comparison, very large, comprising hundreds or thousands of atoms. The action of most medicines is the result of the small molecule of a medicine interacting with a large molecule in the body – often a protein. But, as the models indicate, each type of medicine molecule has a unique shape; it will only interact with a large molecule (called a 'receptor') in the tissues of the body if the shape of the receptor fits it. If it does, it will click into place and slightly alter the shape of the receptor. If it doesn't fit, it will simply pass on by. Figure 1.5 shows how a small molecule, such as a drug or hormone (in bright colours), can fit into an appropriate niche in the surface of a much larger protein molecule (the rough background).

The shape of the drug or hormone and the niche (or 'active site') on the receptor molecule is one important factor determining whether drug and receptor will bind together; another important factor is the way in which electric charge is distributed across both molecules. The two are more likely to stick together if a positively charged zone on the

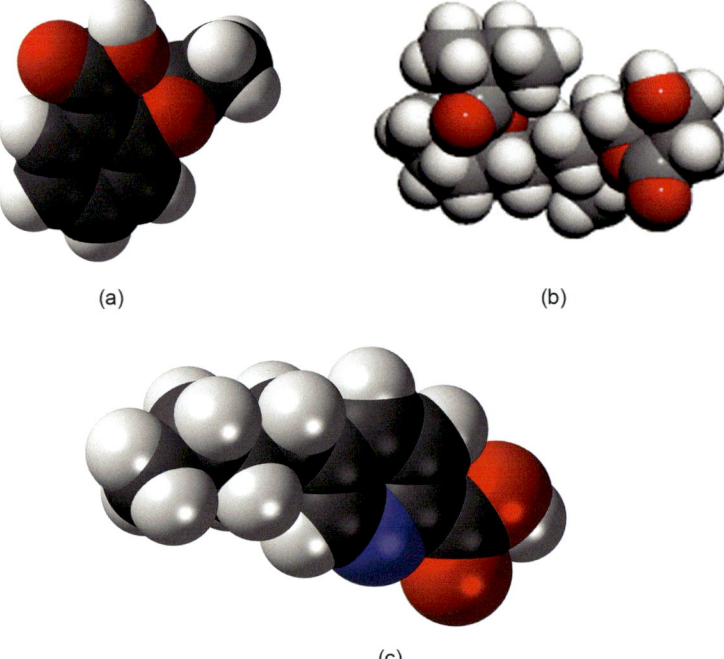

(a)

(b)

(c)

Figure 1.4 Models of molecules of three medicines.
(a) aspirin [Image credit: Benjah-bmm27]
(b) simvastatin (a statin) [Image credit: Fuse809]
(c) fusaric acid (an antibiotic) [Image credit: Jynto].

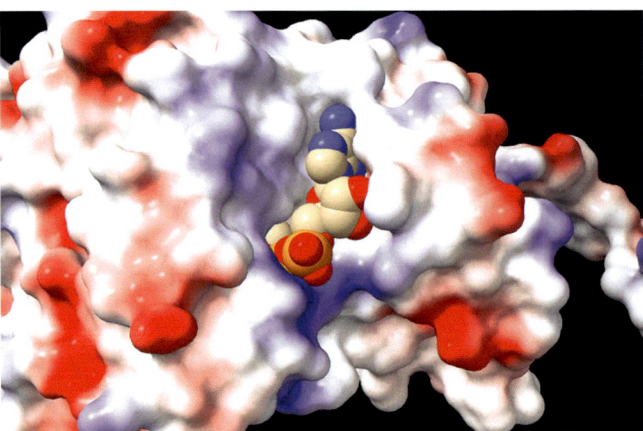

Figure 1.5 Model of a small molecule interacting with a large receptor.
(Image credit: amitkessel.)

drug molecule corresponds with a negative zone on the receptor (or vice versa).

We can now see that a pill only works where the molecules of its active ingredient meet up with a receptor molecule with which it can interact. When you swallow a pill, suck an inhaler or rub in an ointment, the molecules of the medicine course throughout the body, passing by the vast majority of locations, only stopping to interact with the body at specific sites where they match up with the surface of a receptor molecule. Aspirin and ibuprofen molecules, for example, dock onto an active site on an enzyme that produces the pain chemicals, prostaglandins, thereby inhibiting production of these pain chemicals. Statins like simvastatin also act by inhibiting an enzyme, in this case one that helps the body make cholesterol.

In the past, molecules were often developed as medicines by making an informed guess about suitable candidates and then testing them on living systems. Folk knowledge about medicinal plants often acted as a starting point. The antimalarial drug quinine, for example, was developed after it was noticed that Peruvians were able to avoid malaria by using bark from the 'quina-quina' tree. New drugs are often still developed from compounds found in plants and other living things. An example is an anti-cancer chemical called vincristine, derived from the periwinkle flower and used to treat leukaemia.

In recent times, as the precise structures of some of the receptors in the body are becoming known, it is proving possible to design some drugs from scratch. A molecule is created in the lab specially to match the active site on a specific receptor molecule, much as a key matches a lock. Examples include drugs developed to combat HIV-AIDS. These were designed specifically to fit into the active site of an enzyme in the HIV and inhibit it, thus preventing the virus from making copies of itself.

Beta-blockers, as their name suggests, are molecules chosen specifically to block the active sites of receptor molecules, in this case ones in the heart that would otherwise respond to adrenaline. In a similar manner, antihistamine molecules block the sites on receptors in the nose, lungs and other parts of the body that would otherwise respond to histamine. For people with allergies, this interference in the release of histamine from specialised storage cells called 'mast cells' reduces their allergic symptoms.

Rapidly growing knowledge of the size and shape of molecules in the body in recent years has led to formidable progress in the design of new drugs. The process of developing a molecule from a prototype on the laboratory bench into a safe and marketable medicine, however, is challenging and full of pitfalls. There is more to contend with than simply choosing an ideally shaped molecule to interact with a receptor molecule.

Drugs have to be able to pass through membranes in the body – in the digestive tract, liver, lungs, skin or other parts of the body – without breaking up or becoming deactivated. They have to be capable of entering the bloodstream so they get distributed throughout the body. On top of all this, they must neither be toxic nor have unwanted effects on other cells in the body and they must survive long enough as they pass through the organs of the body before being broken down. It's no wonder drug development is such a long and arduous process.

So, returning to our original question: the fact is that the pills we swallow have no route plan as they enter our mouth, skin or lungs. They simply pass through membranes in various parts of the body, pass into the bloodstream (or other systems) and circulate around the body until they happen to pass close enough to a suitable receptor molecule in the target organ. There they interfere with the functioning of the receptor molecule (whether by activating or blocking it) and, provided they've been well designed, improve some aspect of our wellbeing – relieving pain, destroying bacteria or evening out irregularities in the flow of vital substances.

Two

Getting acquainted with birds seems to be ever more popular as smart-phone technology advances. All-singing, all-dancing bird-spotting apps now help us identify species not only by their plumage and habitat but even by their song. Citizen science projects across the world are also attracting ornithological enthusiasts, helping populations to be counted and threats to be identified. Beyond the immediate fun of learning the names of the species or observing their behaviour at the garden feeder, curiosity is easily aroused about deeper aspects of avian culture. Why do they sing? Do they all sing? Why do they flock together? How do they find their way around – especially when migrating over thousands of miles? This chapter explores some of these questions.

BIRDSONG

Not all species of bird sing, but a large number do. Over 5,000 different species of singing birds (oscines) are known. They are able to make their various utterances thanks to one or two membranes that vibrate in a part of their throat known delightfully as the 'syrinx' – named after a nymph in Greek mythology who was transformed into a reed when chased by the god Pan. The reeds made a flute-like sound – the sound of panpipes.

But, of course, not all birds' sounds are melodious, many are abrupt or unmusical. Ornithologists break them into two broad groups: 'calls', which are brief (as in 'cheep-cheep') and 'songs', which are longer and more melodious. As you might expect, these serve different purposes, but both are critical for survival and reproduction. That's how they came to evolve. Calls are commonly associated with alarms: warning of nearby predators, instructing the young or guiding a flock, for example. Small birds may create a chorus of calls around a predator in order to point it out to others and hopefully scare it away. Calls to chicks may command them to stay still when a predator is around, or to swim or peck at food. Chicks, in turn, may use calls to encourage their parents to feed them.

DOI: 10.1201/9781003272779-3

Songs, on the other hand, serve different purposes. Primary amongst these is finding a mate. Singing may attract a mate and can kick off courtship behaviour by signalling readiness to breed. It may also be used between an established pair to help maintain their bond. Research suggests that the quality of singing and the repertoire are associated with the health of the individual. Genes favouring good singing are, as a result, more likely to be passed on to the next generation. This is how birdsong would have been favoured by evolution, bequeathing us the songsters we enjoy today.

Birdsong can also serve a territorial purpose – and yes, it's usually the males that do most of the singing. By getting to a breeding area first and chanting loudly and clearly an individual can mark out and maintain his territory. Sometimes a kind of dialogue with a nearby rival may develop during which boundaries are in effect negotiated. Superior vocal quality can mean that one individual gets established as the dominant one without the need for physical conflict – another way in which the genes of sonorous individuals may survive longer, to be passed on more abundantly.

Each species has its own identifiable song (though, just to confuse us, some species go out of their way to mimic others). The precise pattern can vary according to the age of the individual and the time of year. More surprising, perhaps, is evidence of distinct regional dialects. Different populations of the same species, isolated from one another, may develop distinctive variations on a common theme. Song types have also developed in response to the local habitat. Those that live in dense woodland or reed beds, for example, tend to make louder sounds. An audio clip of the booming sound of a reed-loving bittern is available online (https://commons.wikimedia.org/wiki/File:Botaurus_stellaris_-_Eurasian_Bittern_XC362697.mp3) (recorded by Miro Demko). Other species choose high wires or posts to ensure their chants are heard far and wide. An example of this – the European robin – can be heard in an audio clip (https://commons.wikimedia.org/wiki/File:Erithacus_rubecula_-_European_Robin_XC542842.mp3) (recorded by Benoît Van Hecke).

It's in early morning and evening that most birds choose to sing. An audio clip (https://commons.wikimedia.org/wiki/File:Bourne_woods_Birdsong_and_rain_2020-06-17_0748.mp3) recorded by Bob Harvey in a wood in Lincolnshire is a reminder of that glorious sound. The light level guides them, so a cloudy morning will affect their timing. Because each species reacts to a different level of light, the order in which singers join the dawn chorus follows the gradual brightening of the day. Springtime marks the climax of the singing year as territories and courting are being negotiated. The UK Royal Society for the Protection of Birds (RSPB) suggests that the low levels of light at dawn and evening are favoured as it's too dim to go out hunting for food yet a good time to risk singing out for a mate, while remaining obscure to predators. Things quieten down after eggs have been laid, to avoid signalling their location.

All three audio clips are licensed under the Creative Commons Attribution-Share Alike 4.0 International licence: https://creativecommons.org/licenses/by-sa/4.0/deed.en

A key interest of researchers has been whether bird songs are learned or inherited. An experiment with young chaffinches reared away from all chaffinch song resulted in a very limited kind of singing. This suggests that the ability is partly genetic and partly learned from parents – like so many of our human traits. Calls, on the other hand, with their much simpler patterns, are simply inherited through the genes, rather than acquired through learning.

Like us and other animals, birds produce their sounds where the throat joins the windpipe (or trachea). They don't produce their sound in the larynx part of the throat as we do, however, but lower down, at the point where the two bronchi branch off into the separate lungs. Here the walls are made of flexible membranes for a short distance, and these vibrate as air passes over them. In this area – the syrinx (Figure 2.1) – muscles enable the membranes to tense-up or relax, resulting in a variety of sounds when air flows past them. Vibrations in a medium such as air are the basis of all sound, and structures that vibrate when air passes over them are the basis of many musical instruments.

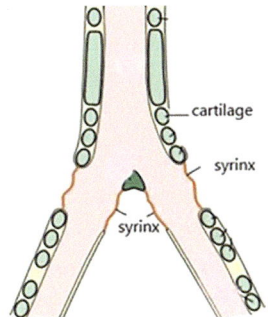

Figure 2.1 Diagram of the syrinx.
(Image credit: Uwe Gille via Wikimedia https://creativecommons.org/licenses/by-sa/4.0/deed.en)

It's the ability of muscles to alter the quality of vibrations in birds and humans (and other species) that enables such a wide range of sounds to be uttered. The location of the syrinx gives birds one special advantage over us humans. With two sets of vibrating membranes, one in each bronchial tube, it is possible for a bird to produce two sounds simultaneously, whereas our single larynx confines us to just one.

MIGRATION

The arrival of Spring, with its warmer weather and longer days, brings out the best of birdsong – whether to declare ownership of a territorial patch or to proclaim readiness for mating. But how did the migratory birds manage to find their way back to their breeding grounds after wintering in warmer regions. And how do they manage to stay aloft for so long while doing so?

Evolution has ensured that bird anatomy is well adapted for flight. Wings are shaped like aerofoils to maximise lift as air streams over them and bones are hollow to minimise their weight. By flapping their wings, birds are able to produce both an upward force to counter gravity and a forward thrust to move them ahead. When conditions are right, air currents move upwards; warm air at the Earth's surface is less dense than cooler air above, so rises. These thermal currents enable birds to soar and glide, much as paragliders do near cliff edges.

A study in Switzerland followed swifts in their migration and showed that they remain aloft throughout their entire stay in Africa. They didn't touch down for a full 200 days. The reason for this and the means of achieving it are not fully understood, but a suggestion is that by staying aloft they avoid predators and perhaps disease too. The energy to remain airborne comes from their diet of airborne insects. It's not certain whether they actually sleep on the wing but they must at least be able to rest in flight if they spend all day and night on the move.

The method of navigation has been studied in many clever experiments and it is clear that several distinct mechanisms are used, sometimes by the same bird. In one experiment, homing pigeons flew straight back home without a problem after having been transported far away and placed in closed, airtight cylinders surrounded by magnetic coils on a tilting turntable! Despite being blasted with random noise and randomly intermittent light, the experiment proved they must have both a compass and a map built in.

Many other experiments have been conducted in an effort to understand how birds manage these extraordinary feats of navigation. There

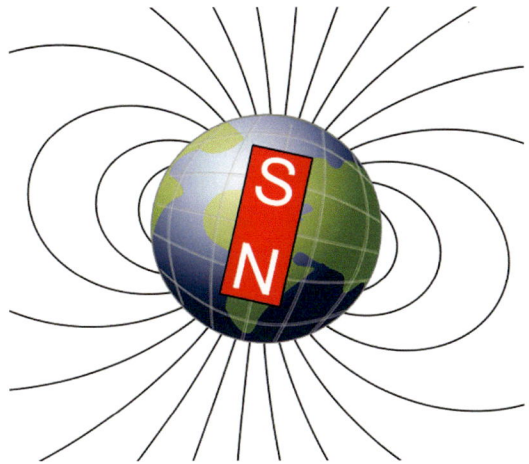

Figure 2.2 The Earth's magnetic field.
(Image credit: Zureks via Wikimedia (Public domain).)

seem to be several distinct methods used by various species. One is by detecting tiny variations in the Earth's magnetic field. As Figure 2.2 shows, towards the north and south poles, the magnetic field gets stronger, as indicated by the closeness of the lines. Around the equator, it is weaker. This offers an approximate sense of latitude.

Experiments at Lund University in Sweden and Oldenburg in Germany suggest that two distinct structures in a bird are sensitive to variations in the strength of the Earth's magnetic field: one acts like a map, indicating where the bird is; the other more like a compass, showing where it is headed. Tiny amounts of the magnetic material iron oxide are stored in the upper part of the beak and these line up with the Earth's magnetic field. These may remind you of the iron filings you may have played with at school, lining up around a magnet to indicate the direction of the magnetic field. This effect in the beak causes cells in the region to send off a signal along the nerves to the brain. Researchers have tested this hypothesis by numbing these nerves and have found that the ability is lost as a consequence. It is thought that slight variations in the strength of the Earth's magnetic field may be detected in this way, creating a kind of mental map in the brain.

The sense of direction – the 'compass' – seems to come from the action of particular proteins in the retina (light-sensitive tissue at the back of the eye). Studies with robins show that when blue light shines on the retina, it knocks an electron out of certain atoms in the protein, altering the magnetic properties of the protein and rendering it sensitive to magnetic fields. This in turn results in a signal being sent along the optic nerve to the brain. It's almost as though a bird can 'see' a magnetic field. Amazing! Through this effect, birds – robins at least – seem to be sensitive to the angle the magnetic field makes with the surface of the Earth – the so-called angle of dip. A glance at the previous diagram shows that this angle is zero at the equator and nearly 90° at the poles – so it gives a clear indication of how far from the equator a bird is – its latitude.

Magnetic cues are just one way in which birds find their way across the globe. The shifting location of the Sun provides another option. In a series of experiments in 1951, starlings were placed in cages in which mirrors were used to shift the direction from which the Sun appeared to shine. As a result, the birds shifted the direction they intended to take off to match the new position of the Sun. As the Sun changes position with the time of day as well as with viewing position round the globe, birds have to be aware of the time of day to work out their location. It appears that the bird's 'body clock' or circadian rhythms enable it to allow for changes in the Sun's position over the course of the day. It seems from experiments with pigeons that young birds have to learn this skill – it's not inherited for their parents. What an extraordinary feat!

A completely different source of information for a bird migrating at night-time is the stars. In experiments involving Indigo Buntings, young birds that had been raised in a lab were placed inside a planetary dome onto which an image of the night sky was projected. This revealed that the buntings did indeed navigate by starlight but did not use the pattern of stars as you might expect. We are accustomed to the illusion of a relatively static backdrop of stars in the night sky, grouped in apparent patterns we call constellations. A first guess might be that birds somehow memorise these patterns and fly at particular angle to them. But experiments reveal this to be wrong. Birds turn out to be smarter than that. In reality, the night sky does not remain static from our point of view on Earth. We are spinning around once every 24 hours, so through the course of a night the stars in the sky appear to us, stuck as we are on the surface of the Earth, to move slowly around a circle (or part thereof, depending on where we are on the Earth's surface). We are usually unaware of this as it happens so slowly and we are not normally prone to staring at the sky for several hours of a night. It takes a long-exposure photograph, such as that shown in Figure 2.3, to reveal this nightly motion.

Figure 2.3 Long-exposure photograph of the heavens from the Earth. (Image credit: Steve Ryan via Wikimedia, https://creativecommons.org/licenses/by-sa/2.0/deed.en)

By good fortune, however, in the northern hemisphere, one star appears to lie at the very centre of these circles – the North Star or Polaris – implying that it must lie directly above the north pole, in line with the axis of the Earth. A short video clip, showing the constellations appearing to revolve around the North Star, is available online (https://www.youtube.com/watch?v=0TFZDO8KcTw).

Birds appear to register the slow rotation of the stars during the night and from this are able to locate the still centre of rotation – the North Star. This enables them to identify the direction of north wherever they are, so they can just point the opposite way to head south (if they are in the northern hemisphere) for their winter migration.

In addition to these remarkable magnetic and celestial mechanisms for finding their way around the globe, some birds, like aircraft pilots, seem to use a more terrestrial method. By monitoring the landscape over which they fly, it appears some species are able to pick out lengthy landmarks such as coastlines, mountain chains or long rivers to follow and, by learning and memorisation, acquire a route map for their migration.

FLOCKING

A gaggle of geese crouching in a meadow, a mass of swifts whirling around a tower, a flurry of long-tailed tits scurrying along a hedgerow – everyday

Figure 2.4 A murmuration of starlings.
(Image credit: Walter Baxter via Wikimedia, https://creativecommons.org/licenses/by-sa/2.0/deed.en)

evidence of yet another aspect of avian culture: their social behaviour. Perhaps the most dramatic example is the giant murmurations created by starlings when they gather in their thousands displaying remarkable aerobatic skill. Figure 2.4 captures an example of this; the movement can be seen in a short online video clip, available in the Support Materials. This video is by Karen Roe, licensed under the Creative Commons Attribution 2.0 Generic licence: https://creativecommons.org/licenses/by/2.0/deed.en

But what's it all about? How does the individual benefit from these kinds of group activity? The UK Royal Society for the Protection of Birds has brought together the results of research on the flocking habits of many species. As with us humans there are several distinct advantages to social activity. A simple starting point is the obvious advantage that hundreds or even thousands of pairs of eyes offer when looking out for potential predators. It's safer for the individual when lots of others are on the lookout. Grouped tightly together, individuals can also move around in a dazzling display to confuse a predator. Mobbing them as a pack can be even more effective. From the predator's point of view, it can be hard to pick out a single bird when surrounded by so many.

Another reason for working in groups is to hunt for food. Geese, for example, often forage together, enabling all to benefit from whichever one finds the source. Some species take it further by nesting close together. Keeping warm in winter is another advantage of close living. Small birds will sometimes share a small space for this very reason.

In flight, flocking also offers significant energy saving advantages in much the same way as a peloton does for racing cyclists. By sticking close behind a front-running individual, a trailing bird (or cyclist) can enjoy the calm of the slipstream left behind by their front-running fellow. Less energy is used up in overcoming the viscous drag of the rushing air, making a longer journey possible. The V formation of a skein of geese is a special case of group benefit (Figure 2.5). In this case, each bird

Figure 2.5 A skein of geese.
(Image credit: Mike Prince via Wikimedia, https://creativecommons.org/licenses/by/2.0/deed.en)

Figure 2.6 Vortex at aircraft wing tip.
(Image credit: NASA.)

positions itself carefully in relation to the tip of the wing of the one in front. In the right position, they benefit from an updraft resulting from the vortex produced at the wing tip as air rushes over it.

The powerful effect of this is illustrated in a dramatic photograph from NASA research (Figure 2.6). Disturbance to the air at the tip of an aeroplane's wings causes the air to rotate in a huge horizontal vortex. The coloured substance injected into the air in this experimental study helps us visualise where the air is beating down on a wing and where on the opposite side of the vortex it provides an updraft.

CONCLUSION

Once again, exploration of matters of everyday interest – the fascinating behaviour of birds – leads us into diverse branches of science: from animal behaviour and throat anatomy to magnetism and astronomy. The effort and imagination of countless scientists and nature lovers around the world help illuminate the everyday world we observe.

Three

I was leaving my doctor's surgery on a bright sunny day when a brilliant rainbow of coloured light on the surface of the floor arrested me. I stopped, amazed by the brilliance of the spectrum projected by a beam of sunlight, and took the photograph in Figure 3.1. Showing it to some friends later prompted them to question what exactly was happening here.

The light from the Sun seemed to be catching the edge of a glass door. It was as though it was acting as a kind of prism – one of those triangular shaped pieces of glass you may once have played with in a science lab at school. What is it about sunlight that produces such colourful effects, they asked?

THE NATURE OF LIGHT

Light has always fascinated people: artists, philosophers, scientists and ordinary folk alike. In early medieval times, it was thought to be an emanation from a kind of fire in our eyes. From around the thirteenth century, it began to be understood the other way round, as something emerging from objects themselves rather than our eyes. Some objects produce their own light – stars, the Sun, lamps and candles, for example – while others, such as people, chairs and the Moon, reflect whatever light shines on them. Today, we understand that light carries energy from its source to the eye, where it stimulates the retina, leading to the creation of a mental image in the brain.

Newton, in the seventeenth century, saw light as a stream of 'corpuscles' or particles flowing from luminous objects. Experiments by Thomas Young the following century indicated that it was more like a rippling wave, apparently contradicting Newton's theory. The idea of light moving as a wave seemed strange, difficult to imagine – as it must surely still be for most of us today. If light did indeed travel as a wave, it was assumed, by analogy with water waves, there must be some kind of invisible substance permeating all space in which these waves occur. This mysterious

DOI: 10.1201/9781003272779-4

Figure 3.1 Spectrum of sunlight.

medium, dubbed the 'luminiferous aether', couldn't be seen or sensed in any other way. With the advance of understanding of electrical and magnetic influences during the nineteenth century, these waves were interpreted as incredibly rapid oscillations of tiny electrical and magnetic fields. These vibrations are set up by the source of light – the Sun, a light bulb, a fire -– and then travel outwards at an unimaginably high speed – the 'speed of light', just over a billion kilometres per hour.

In 1887, two American physicists, Albert Michelson and Edward Morley, set up an experiment to study the so-called aether. They shone a beam of light forward in the direction of the motion of the Earth and another beam at right angles to this. The unexpected result was that the beam of light hurtling forward through space with the additional speed of the Earth didn't travel any faster than the other one, travelling sideways on, perpendicular to the path of the Earth. This scotched the whole idea that the aether existed. The inevitable conclusion was that light did not travel as undulations in a physical medium at all. Light from the Sun and the stars travels through empty space, a vacuum, as an electromagnetic disturbance requiring no medium to support it – not easy to imagine, then or now.

The wave model conceives of light as an immaterial vibration. What vibrates is not a physical thing, but a tiny, tiny electrical influence, a micro voltage, rising and falling, then reversing direction. Accompanying this, a similarly oscillating magnetic influence rises and falls at right angles to the electrical one. To capture their electrical and magnetic nature, these waves are called 'electromagnetic'. Like ripples on a pond, these electromagnetic oscillations move outwards from their source. Figure 3.2 indicates a magnetic field that rises and falls (B, in red) and an electric field that rises and falls with it, but in a perpendicular direction (E, in blue).

A helpful animation of this process can be seen at https://upload. wikimedia.org/wikipedia/commons/9/99/EM-Wave.gif

The strength of these oscillations is miniscule, but it's nevertheless enough to activate the molecules in our retina that give us vision. Unlike waves on a pond, electromagnetic waves don't need water, glass, air or any other medium to support them. Difficult though it is to imagine, they simply travel onwards relentlessly through the void of space, conveying the energy in light – and related forms of radiation, such as UV and infrared – from distant stars. It is this light, bouncing off the molecules in the atmosphere, that gives us daylight and, reflected off the objects around us, enables us to see at all.

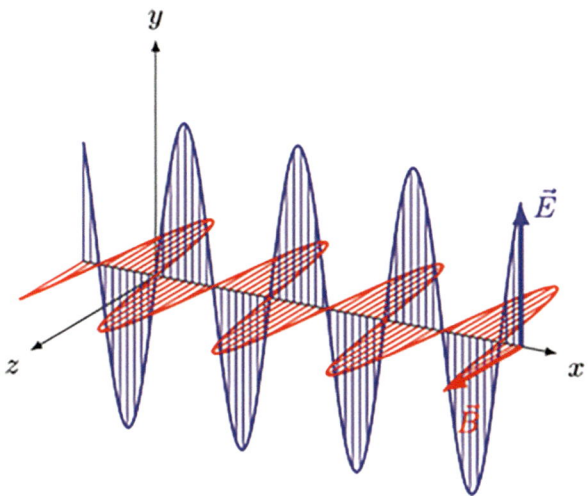

Figure 3.2 Electromagnetic wave.
(Image credit: And1mu via Wikimedia.)

A subsequent revelation about the motion of light in a vacuum resulted in a radical change in our understanding of physics. It had been discovered that, when light shines on a metal, it could, under certain conditions, cause an electric current to flow in the metal. The effect of varying the brightness and the colour of the light had been studied, the anomalous results of which could not be explained by the prevailing wave model of light. Einstein proposed, in a famous 1905 paper, that the light was not behaving as an undulating wave in this case, as expected, but instead as a stream of particles, akin to Newton's obsolete corpuscular model. The energy of these particles is related to the frequency of the light: higher frequency light corresponding to higher energy particles. If light was assumed to act like a stream of discrete particles (dubbed photons) in this situation, the experimental results could be explained. At the same time, however, other experiments continued to imply that light behaves as a wave.

Physicists grappled with this dilemma, seeking ways to reconcile the apparent contradiction, without success. Today, we have come to accept that both theories are valid in so far as they explain the findings of different kinds of experiment. In practice, we alternate between these apparently incompatible models to explain different aspects of the behaviour of light. Generally, the photon model best describes the emission of light from a source and its absorption, while the wave model best describes its journey from source to detector. Physics is pragmatic about this impasse; the essential nature of reality remains a question for philosophy.

COLOUR

As we know from everyday experience, vibrations can be rapid or leisurely. A pendulum clock oscillates at a relatively slow pace, whereas the wings of a humming bird move up and down far more rapidly. In more precise language, the frequency can be low or high (Figure 3.3). As you might expect, the more frequent the vibrations, the shorter the distance between peaks, known as the wavelength.

The Sun, or any other hot source, is a source of electromagnetic waves of a wide range of frequencies. Our eyes respond to some of these frequencies, but not others. What our bodies recognise as visible light are waves whose frequencies lie within a certain range. Waves of frequencies outside the visible range are not picked up by our eyes but may be detected by other means. For example, we experience one type of lower frequency waves (known as infrared) as heat; still lower frequency

Low frequency

High frequency

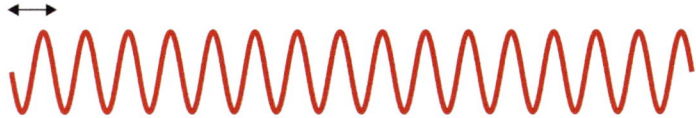

Figure 3.3 Frequency.
(Image credit: opentextbc.)

radio waves are not detectable by us humans, but are picked up by radio aerials or antenna. An animation showing this can be seen at https://upload.wikimedia.org/wikipedia/commons/b/b7/Dipole_receiving_antenna_animation_2_800x403x15ms.gif

Higher frequency waves such as ultraviolet (UV) or X-rays can be detected by their effects on, for example, photographic plates. We humans aren't able to sense these, but they can damage our tissues; that's why we need to avoid exposing ourselves too long to strong sunlight or X-rays in hospital.

Within the visible range, our eyes are capable of not only *detecting* light waves but also of differentiating their various frequencies. Different molecules in the cells of our retina respond to low, medium and high frequency light waves. These are known as red-, green- and blue-sensing molecules respectively as they respond most strongly to these colours. Signals are sent from the eye to the brain via the optic nerve, giving detailed information about light entering the eye, including the various frequencies included in it. The most extraordinary transformation happens next: our brain associates the various frequencies of the light impinging on our retina with what we sense as colour. The human sensation of colour is a creation of our brains. It's a philosophical question whether colour actually exists in the external world, given that it's brains that create the sensation. As Sonya once remarked in a discussion: 'if I put my jumper in a cupboard is it still red?'

The beautiful rainbow colours cast by sunlight catching the edge of a glass door arise from a spreading out of the light of differing

Figure 3.4 Merging of colours in a spectrum.

frequencies. It's a reminder that light from the Sun is not of one pure colour (monochromatic), but a composite of waves of differing frequencies. As you see in any spectrum or rainbow, the spread of colours is continuous; there's no border defining the end of red and beginning of orange. It's a continuum of gradually shifting hues, as the photograph in Figure 3.4 confirms. Our quest is to understand what it is about the Sun that makes it radiate light over this continuous range of wavelengths.

Light originates in atoms: more precisely, in changes that occur within atoms. Under normal conditions here on Earth, gases, when heated to high temperatures, emit a strictly limited range of frequencies of light. A good example is the ubiquitous orange street lamp, which contains sodium vapour. When heated this emits light of a single frequency and hence colour: orange. In contrast, light from the Sun comes from its very hot surface layers, where atoms and freely moving particles (called electrons) are buzzing around at very high speeds and interacting with one another frequently. This causes the atoms to emit light which, under these extreme conditions, is of a very wide and continuous range of frequencies. As a result, the daylight we see includes all these frequencies – i.e. all the colours of the rainbow. In addition, there are also frequencies that lie outside the range of visible light: UV which has a higher frequency than violet light and infrared which has a lower frequency than red light.

The jagged white line in Figure 3.5 shows the intensity of the Sun's radiation over a range of different wavelengths. The intensity is depicted

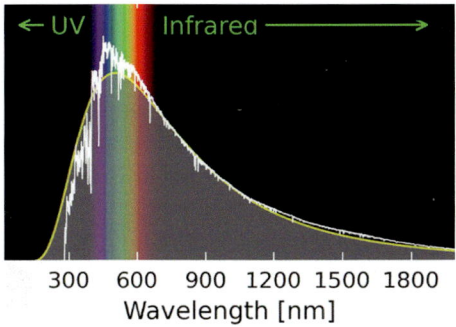

300 600 900 1200 1500 1800
Wavelength [nm]

Figure 3.5 Intensity of sunlight as a function of wavelength.
(Image credit: Danmichaelo via Wikimedia (public domain).)

along the vertical axis and the wavelength along the horizontal one. You can see that the peak of the white line in the graph (i.e. the greatest intensity) occurs for visible light, where the spread of colours from violet to red is shown. Much of the radiation from the Sun is what we see as light, in all its colours, but a considerable fraction, is beyond the visible: in the infrared range (which we experience as heat) and the ultraviolet (UV).

THE SPECTRUM

When all the colours of the spectrum enter our eyes together, our brain represents this as what we know as white. Like all colour impressions, this is an artefact of the brain. The Sun, however, does not actually look white from here on Earth, despite the range of colours it radiates. It appears, instead, in different colours at different times of day. This is not because the Sun actually changes colour, but as a result of its light passing through the Earth's atmosphere before it reaches us. Light waves interact with molecules and particles in the atmosphere and are scattered by them, to some extent, in all directions. The various frequencies or colours in the light are, however, scattered to different degrees. The bluer frequencies are scattered more strongly, upsetting the balance that makes for an overall white colour. The Sun appears closest to white when it passes through the least depth of atmosphere, i.e. when it is overhead at noon. In the morning and evening, when sunlight reaches us at a glancing angle, it passes through a longer stretch of the atmosphere and, as a result, loses more of its blue component. With yellows and reds becoming relatively stronger in the mix, the Sun appears more yellow or orange away from its noonday peak.

The final step in our query is to explain why the great mixture of frequencies in the light from the Sun should get split up into the colours of the rainbow by a chance encounter with a glass door. In essence, we are asking why light gets split into its constituent colours when it passes through a prism. The edge of the glass door, held ajar, with the sunlight playing upon it at a glancing angle, was acting just like a prism.

This spreading out of the frequencies or colours is a consequence of light being refracted: the wavefront bending as it hits a boundary. The underlying reason for this is that light travels at a slightly different speed when passing through a physical medium rather than a vacuum. In Figure 3.6, the wavefronts are travelling at an angle to an area of water or glass that lies ahead. Light waves travel more slowly in glass or water than in air, so after hitting the boundary the wavefronts move at a slower pace. The effect is that the front of the wave changes direction.

A helpful animated version of this image can be seen at https://commons.wikimedia.org/wiki/File:Refraction_animation.gif

In the case of a prism, however, the light is bent twice in the same sense, both as it enters and leaves the prism.

Having seen why light gets bent as it passes through a prism, it remains to be explained why the colours in sunlight get separated out in the process, as shown by the spectrum that opened this chapter, caused by sunlight catching the edge of a glass door. As we have seen, sunlight includes waves with a spread of frequencies or colours. Light of all these frequencies slows down when it enters a medium such as glass. However, each frequency or colour slows down to a slightly different degree. Red light slows down least and violet the most. Figure 3.7 shows each

Figure 3.6 Refraction at a boundary.
(Image credit: Ulflund via Wikimedia.)

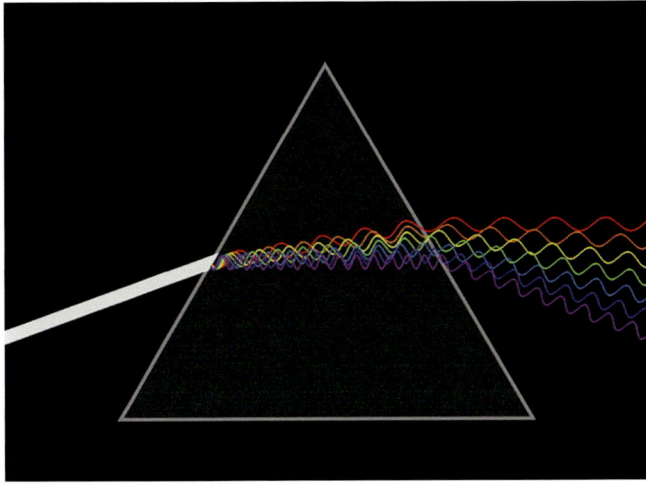

Figure 3.7 Dispersion of white light in a prism.
(Image credit: Lucas Vieira via Wikimedia.)

colour bending (or refracting) by a slightly different amount as it hits
the glass at an angle and also as it leaves. The sunlight not only bends as
a whole as it passes through a prism – or the edge of glass door – but, in
addition, separates into its hidden range of colours: the spectrum.

An animated version of the image can be seen at https://upload.wiki
media.org/wikipedia/commons/f/f5/Light_dispersion_conceptual_
waves.gif

The arresting sight of sunlight on a bright day spreading its colours
across a shop floor has led us to explore some of the most fundamental
ideas in physics: the nature of light, the meaning of colour, the interaction
of light with solid matter – a beautiful sight with beautiful theories to
explain it.

Four

'Who's for coffee?' – a familiar enough question at the end of a meal, but not usually one that opens up discussion of the genetics of inheritance. As a family group was finishing lunch one afternoon, one of the brothers turned down the offer, claiming it interfered with his sleep; the other chose an espresso, claiming it had no such effect on him.

'Ah – I've read about this – caffeine – it's genetic' chipped in the sister. 'How can it be, if they are brothers with the same genes?' queried another. 'Hang on', added a friend, 'don't you get some from your mum and some from your dad? You get a mixture, don't you – one may be blue-eyed, the other brown?'

The confusion, with its half-remembered ideas about how inheritance works, was not surprising. It's not easy to fathom out why some siblings look similar and share some of their parents' traits and others don't. Even more baffling is how, despite inheriting our genes exclusively from our parents, we manage – every single one of us – to be a unique human being. This is the story of inheritance, and the genes that make it happen.

THE CASE OF CAFFEINE

Let's start with caffeine. It turns out that there is indeed a genetic component to our love of coffee … and indeed our individual response to it. According to a 2010 review of research, somewhere between 36% and 58% of the differences between us when it comes to our responses to caffeine can be put down to our genes. Two genes are involved: one enables caffeine molecules to be broken down, the other controls the extent to which this happens. As we became only too aware during the Covid-19 pandemic, however, each gene comes in various versions, thanks to its capacity to mutate. In the case of our response to caffeine, one version causes caffeine to be broken down rather quickly, so that its effects will fade more rapidly for people who have this version. This is just one of the ways we differ from one another even if we're closely related – different versions of the same gene.

DOI: 10.1201/9781003272779-5

Preferences for coffee may seem a rather obscure way to start a discussion on genetics, but what's true for caffeine goes for almost all our bodily functions and inherited traits. The same intricate mechanism of inheritance, by means of genes, holds throughout our bodies and indeed the entire living world, from the humble yeast to the human brain. We explore here how genes and their molecular partners – the proteins – account for that most universal of interests: how children resemble their parents (or not).

GENES

Eager relatives are keen to check out new-born additions to their families, looking for tell-tale features from their mum or dad, or even grandparents. This is entirely reasonable, given that the path of physical development from the womb onwards is so strongly influenced by information coded from the very beginning in their genes. Each gene carries the information needed to make one of the many thousands (possibly hundreds of thousands) of types of protein in our bodies. And proteins go on to shape the way our bodies (including our minds) develop. They make up the tissues of which our muscles, skin and hair are made and also act as enzymes, activating vital processes such as digestion, breathing, communicating and thinking.

What we become, even as an embryo, let alone as an adult, is largely determined by the actions of the proteins, for which our genes provide the blueprints. To take a single example, we grow tall (or not) in large measure through the action of a protein called human growth hormone which stimulates our bones to grow. The particular gene that provides the code for making this hormone is therefore a key factor in determining our height. Somewhere between 60% and 80% of the influence on how tall we grow is genetic, according to studies of twins and siblings. Of course, other environmental factors, such as nutrition, play their part in shaping how tall we actually become – these constitute the other 20%–40%.

As with very many aspects of our bodily functioning, it is not only one gene and its corresponding protein that determines the height to which we will grow. Many genes are implicated in the complex process of growth and each of these comes in a number of versions. Research has uncovered over 700 different variants of the various genes that help determine our height.

The complexity of pathways by which genes influence how our bodies work has come as something of a surprise in modern times. There

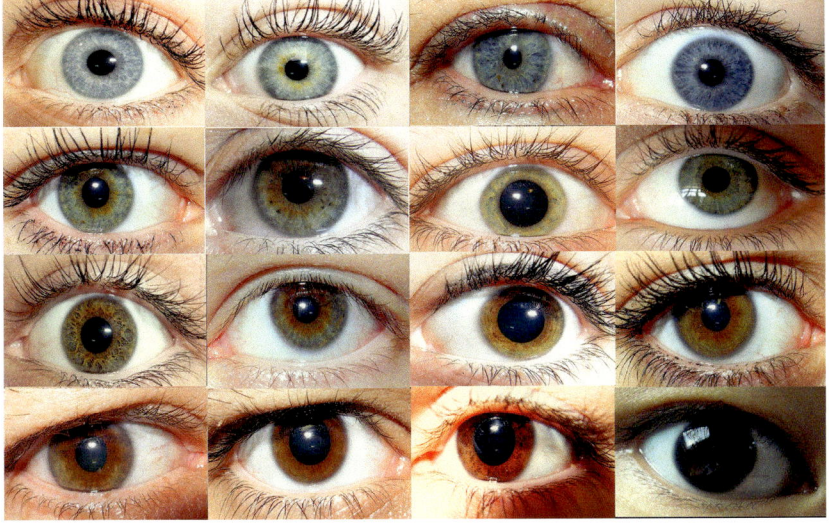

Figure 4.1 Eye colours. (Image credit: LeuschteLampe via Wikimedia (Public Domain).)

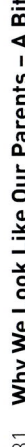

was an earlier expectation that each characteristic of our bodies might be determined by a single gene. This idea has been popularised by the example of eye colour in which a single gene was thought to provide the code for a single protein which stimulated the production of melanin, the pigment that gives the colour to brown eyes. We now know that different shades of eye colour are due to different amounts of melanin in the iris and also to the way in which the melanin molecules interact with the light falling on them. Both of these differ between individuals and, as a result, there is a range of different eye colours, as Figure 4.1 shows.

To add to the complexity, we now know that eye colour is not determined solely by one gene: several different ones play a part, allowing for many possible hues between blue and brown. As research advances, it is becoming increasingly clear that the metabolic processes that shape our physical and mental selves are complex pathways, involving many stages, many proteins and many genes. There's a long way to go till all of these are understood in any detail, and even further till we understand how they interact with environmental factors. Nevertheless, some characteristics have been found to be determined by a change (known as a mutation) in one particular gene and some pathways have been clearly elucidated. Of practical importance is research into the genetic basis of some heritable disorders, including cystic fibrosis, sickle cell anaemia and Huntington's disease (see below how gene therapy is beginning to help with this).

ALTERNATIVE VERSIONS

If some versions of genes are defective, we are bound to wonder why heritable disorders, such as cystic fibrosis, affect some offspring but not others. Why is it that some people carry an unhealthy gene without their children being affected, whereas the children of others are? Rosie, thinking aloud about this issue in a group discussion one evening, wondered whether this might be linked to the taboo on close relatives marrying in many cultures. Inspired by this thought, Julie recalled reading that, in some isolated communities, strangers were particularly welcomed as it held out the prospect of widening their gene pool.

As we all know, we inherit our genes from our biological parents; but, of course, we have two of these, each with a different version of every gene; whose do we get – mum's or dad's? By reproducing sexually, we receive two versions of each gene rather than the single version that clones have to manage with. Over the aeons, evolution has brought about this development, which can offer an advantage if one version of a gene is defective but the other is healthy. In the unlucky situation that both parents carry a faulty gene as one of their pair, there is a significant chance (25%) that both versions their offspring inherit will be faulty. This is the cause of the genetic disorders mentioned above – sickle cell anaemia and cystic fibrosis for example.

GENETIC ROULETTE

The puzzling thing about family resemblance is that we seem to be like our siblings and parents in some respects and quite different in others. Susan, a young mother in one discussion group, expecting her first baby, was intrigued by this question. How is it that two acts of conception between the same parents can lead to children with utterly distinct characteristics? What is happening to the genes that gives us this endless variety?

The statistical trick is that, with two different versions of each gene available, and over 20,000 distinct genes in a human being, the number of permutations is simply enormous. Imagine a coin tossed 20,000 times and try counting the number of different ways it could have fallen. Just toss it twice and you have four ways (HH, HT, TH, TT), toss it three times and you have eight .... imagine 20,000 times; what are the chances of getting 20,000 heads in a row? It's this kind of unlikeliness that makes us each unique. So, the question of what makes you you largely boils down to which version of each gene gets passed on from your parents at the moment of conception. Amazingly, it's not so different from coin tossing: there's only two choices (usually) and chance determines your ultimate make-up. If any one of a number of random events had worked out differently, a different baby would have been born.

The extraordinary story of gene versions (or alleles) and the way they get mixed in each act of reproduction helps us understand why every individual is unique, yet resembles their close relatives. It's not the whole story however. Genes go on to instruct the cells in our bodies which proteins to produce. But many other factors will subsequently determine how plentiful and how successful these proteins are in developing us, right through from tiny embryo to ageing adult. Factors in our nutrition, our behaviour, the air we breathe and the stresses to which we are subjected will shape our physical and mental development.

GENES, CHROMOSOMES AND DNA

We have seen how the gene versions from each parent get mixed in a random fashion to make up the genome of their offspring. But just how is this remarkable feat achieved? To explore this, we need to look at what is going on at the molecular level. A gene is just a short stretch of a long molecule of DNA – the famous double helix. The DNA molecule is extremely long and very thin – very much more so than the thread on a cotton reel. To fit such a long thin threadlike DNA molecule into a microscopic cell, it has to be coiled up like the cable of an old-fashioned telephone. And the coiled structure is itself coiled up in turn, as shown in Figure 4.2.

This supercoiled DNA does not occur in cells as one single long thread, however. Instead, it is divided up amongst 23 distinct structures called chromosomes (46 if you count the ones from each parent). All the 20,000 odd genes we need for our bodies to function are contained in just 23 pairs of distinct chromosomes and these are replicated in every one of the trillions of cells in our bodies. It's the genes in these chromosomes that produce all the proteins that make our bodies work, day in day out, throughout our lives. Genes produce the enzymes that run our metabolism 24/7: the haemoglobin that carries oxygen from our lungs and the proteins that contract our muscles, for example. Genes are in use all the time, throughout our bodies, supplying the code to make our proteins. From time to time in our lives, however, we need to do more than just function: we need to create the next generation!

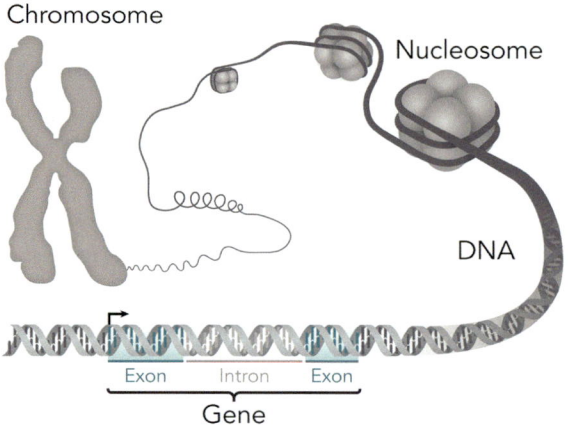

Chromosome

Nucleosome

DNA

Exon Intron Exon

Gene

Figure 4.2 How DNA coils up.
(Image credit: Thomas Splettstoesser via Wikimedia, https://creativecommons.org/licenses/by-sa/4.0/deed.en)

Like the overwhelming majority of animals and plants, we don't do this the simple way by cloning – making exact photocopies of ourselves. Instead, we make unique individuals, based on ourselves, through the process of sexual reproduction. This ensures that variation is built into the way the species evolves. It enables individuals to emerge with differences introduced by chance at the moment of reproduction. A few of these differences might fit the individual better to the environment in which they live, enabling them to survive longer and reproduce more. This is how successful species gradually adapt to their ever-changing surroundings over very many generations.

Despite its complications in the short term, sexual reproduction generally beats cloning in the long run. The complications are less to do with the physical acts of mating or fertilisation, more to do with what has to happen at the cellular level. Somehow reproductive cells from two parents, each equipped with two versions of every gene have to mix their four versions together, yet end up with offspring that have only two versions, one from each parent, just as each parent cell has. This is why very special kinds of cells are necessary: the germ cells that are required to produce the ovum (egg cell) and sperm. These special cells are produced in such a way that, unlike all other cells in the body, they end up with only one version of each gene, instead of the usual two. This complicated halving process – called meiosis – takes place in the ovaries and testes.

The germ cells that will go on to produce the egg cells of a mature woman are already in place in the embryo of a baby girl as she develops in her mother's womb. The ova she will go on to release each month in later years are already present before she is even born (though not yet ready for use). The equivalent germ cells that will go on to produce sperm cells in a mature male are also already in place before birth. It is not until puberty that sperm cells themselves get produced from these forerunners, as a result of hormonal action.

VARIATION

Variety in our genes is, in large measure, what makes us each a unique individual; it is also what enables our species (or any other) to evolve over millions of years, as the environment changes. This comes about because the genes we inherit from our mother and father get mixed randomly as our germ cells are created. In an extraordinary process of re-organisation within each germ cell, the chromosomes that carry the genes from our mother line up next to those from our father in such a way that equivalent genes from each parent lie physically opposite to each other. Research in the past 50 years has shown that, when they line up in this way, adjacent DNA molecules – which are carrying the all-important genes – can cross-over or recombine at various points along their length as shown in Figure 4.3. In this way, the version of a gene from one parent (red) can get 'welded' into the DNA from the other parent (blue). The result is a novel version of each chromosome, containing a mixture of gene versions from each parent (lower part of Figure 4.3). Amazingly, the mixing is random – there is no way to predict which version will get transferred into which chromosome.

The two newly blended chromosomes then go onto the next stage of reproduction in which just one of them gets to enter the egg or sperm cell. Once again, the selection is random. So, there is no way of telling which combination of gene versions (or alleles) will go on to make a new baby.

In fact, the unpredictability is even greater than this. Each germ cell that will go on to produce sperm or egg cells is itself a unique random combination of genes. Every egg and every single sperm cell will be genetically unique. If sperm number one-million-and-two reaches the egg first, at the moment of fertilisation, it will result in a different baby than if sperm number one-million-and-three makes it first. We are each truly unique, and extraordinarily unlikely, individuals.

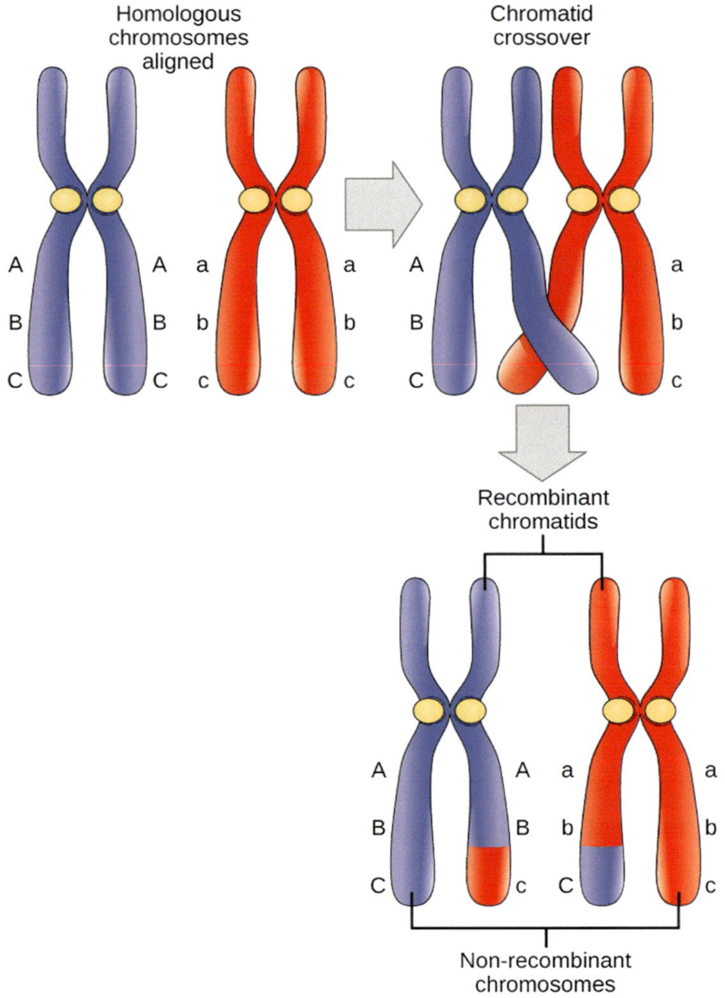

Figure 4.3 Recombination: mixing genes from parental chromosomes. (Image credit: CNX OpenStax via Wikimedia, https://creativecommons.org/licenses/by/4.0/deed.en)

But let us not forget that, despite the extreme chanciness introduced by this mixing of gene versions (alleles), all our genes are inherited exclusively from our parents. The pool from which ours are selected is the same as that of our siblings – it's the particular mix of versions that gets selected that differs. That's why we are not identical to our siblings but may well resemble them. And what is true for us was also true for our parents – they inherited their combination of alleles from their parents and so on throughout our ancestry. This explains how traits from grandparents, or even great grandparents, may appear in later generations – the alleles are passed on largely unaltered from generation to generation.

INHERITED DISORDERS AND GENE THERAPY

As mentioned earlier in the chapter, one version of a gene may, in some cases, be faulty in some way – it may contain an error in the code for making a vital protein, for example. Where the gene version inherited from one parent is healthy, this will usually be dominant over a faulty one, meaning that any offspring are healthy, despite carrying one version of the gene that is faulty. It occasionally happens, however, that both parents carry the faulty version and pass it on, by chance, to a child. In this case, no healthy version of the gene is available to the offspring. This can result in a serious genetic disorder.

An example is sickle cell anaemia. In this case, a particular gene that carries the code for making the haemoglobin protein is abnormal: one letter in the sequence of the code is incorrect. The diagram in Figure 4.4 represents a stretch of DNA from this particular gene. The dark blue region shows a sequence in the code in which the letters C, T and T represent three chemical groups (known as Cytosine, Thymine and Thymine). In the 'normal' situation (left-hand side), this sequence will cause the chemical labelled Glu (glutamate) to be inserted in the nascent protein being built, based on this code. For people with sickle cell anaemia, however, a faulty (mutated) version of the gene contains a letter A (for the chemical adenine) where a letter T should be.

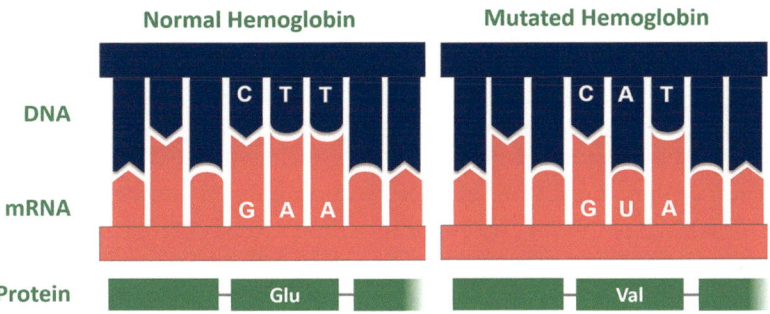

Figure 4.4 Gene mutation in sickle cell anaemia.
(Image credit: Thomas Samuel for ACC-BioinnovationLab via Wikimedia, https://creativecommons.org/licenses/by-sa/4.0/deed.en)

The CAT sequence means that the wrong component – known as Val (valine) – is put into the haemoglobin, giving rise to a malformed molecule. This malformation causes the haemoglobin molecules to clump together, reducing their capacity to carry oxygen and distorting the shape of the cell in which they are contained. Fatigue and severe pain are some of the consequences for sufferers.

In recent years, however, the development of gene therapy offers the prospect of being able to remedy faults in some genes in certain cases. This means the mutation in a patient's genes might either be corrected or, in other cases, the gene completely replaced with a healthy one. This promising area of medical research is still at an early stage for most diseases in which genes play a role; huge technical challenges remain. The first of these is to find a way to produce the healthy genes in a laboratory setting. A way then has to be found to get these into the relevant cells of the body and a method established for replacing all copies of the mutated versions.

Amazing progress has been made on each of these in the past few decades. In the example of sickle cell anaemia, a beneficent virus is used to break into the cells that make red blood cells in the bone marrow. This exploits the special ability of viruses – to get inside healthy cells – but, of course it's done in a carefully controlled way. Once inside, these viruses carry the gene for healthy haemoglobin into the cell. Unwittingly the cell starts producing healthy haemoglobin from these imported genes rather than the less effective ones it already had. Red blood cells are refreshed relatively quickly (a matter of months), so gradually all the defective cells get replaced. Because of the many technical challenges, however, most genetic conditions have not yet proved amenable to gene therapy. Who knows what research will bring in years to come?

CONCLUSION

The extraordinary story of gene versions (alleles) and the way they are brought together in each act of reproduction helps us understand why every individual is unique, yet resembles their close relatives. It's not the whole story however. Genes not only transfer vital information at the moment of reproduction, they also make this information available in the production of the various proteins needed by all the cells of our bodies. Exactly how we go on to develop as individuals over the course of our lives, from tiny embryo to ageing adult, is of course not determined solely by our genetic inheritance. Many other factors will shape our physical and mental development. Trying to work out the interplay between the genetic and environmental influences on our lives and wellbeing is a matter of great interest to all of us. Fortunately, it is also an increasingly important area of scientific and social research, holding out the prospect of better medical treatments and a more nuanced understanding of the human condition.

Five

The dramatic image in Figure 5.1, produced by the US National Weather Service, shows a special weather event bringing huge rain storms to California. You can clearly see the 'river' of high humidity air streaming across from Hawaii to California – known colloquially as the 'Pineapple Express'.

A friend in California got in touch to ask about these so-called atmospheric rivers after witnessing the devastating impact in her area. Similar extreme rainfall events are happening more and more often across the world. 'What causes rain to fall in the first place?' she asked, and 'Why does it seem to be on the increase?' In this story, we explore the chief scientific concepts underpinning why it rains so much – at least in some places, some of the time.

The image in Figure 5.1 reveals how much water vapour is locked up in the atmosphere at every point. The darker blue colours indicate the higher concentrations of water vapour. They show how the warm conditions over Hawaii enable large quantities of water to be absorbed into the atmosphere by evaporation from the oceans and how the wind moves this eastward towards California. Water-laden air moving from the Pacific Ocean eventually makes landfall in California and soon afterwards meets the immovable object of the Sierra Nevada mountains. As Figure 5.2 shows, the air is forced upwards as it crosses the mountains and, in so doing, cools down (it's colder up there). The result is rain, or even snow and, if conditions are bad, this can lead to extreme flooding down in the plain.

But, as this book is about underlying scientific concepts rather than weather forecasting, let's take the opportunity to dig a little deeper and ask what exactly is going on here. What indeed is a vapour? Why should it lead to rain when it cools? Answering these questions opens up a whole world of interesting phenomena – clouds, mist, frost, dew, humidity, perspiration – even the problem of drying your clothes on a wet Wednesday.

DOI: 10.1201/9781003272779-6

Precipitable water forecast for mid-day, April 6, 2018

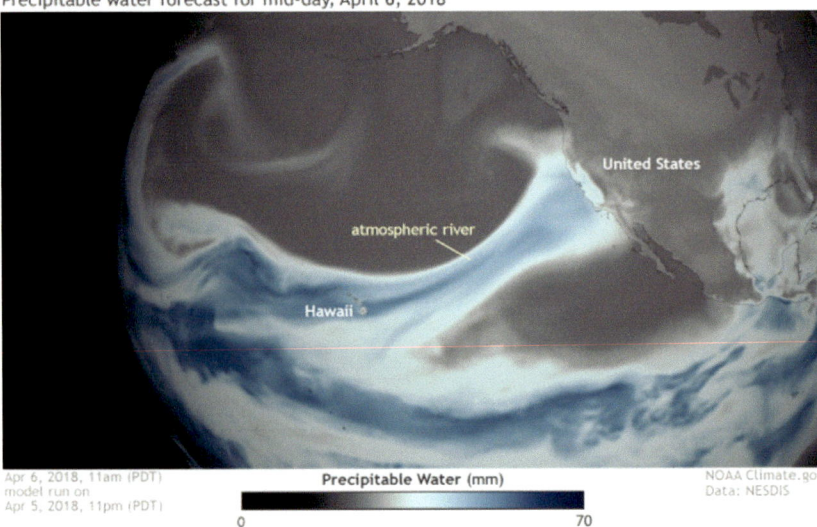

Figure 5.1 Atmospheric river.
(Image credit: US National Weather Service.)

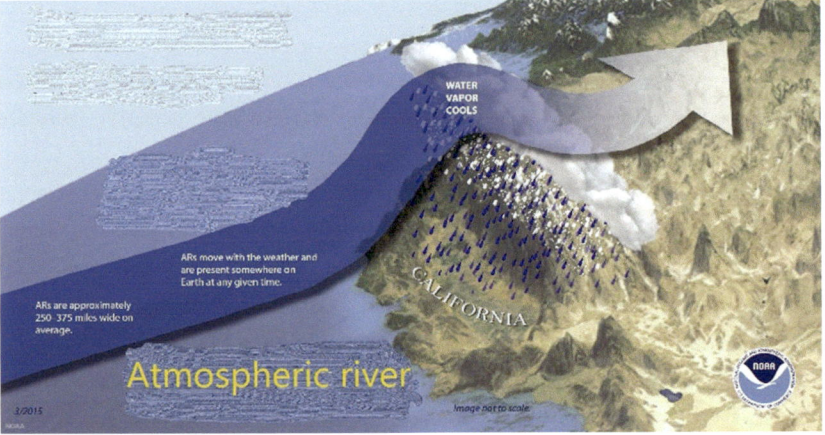

Figure 5.2 The science behind atmospheric rivers.
(Image credit: NOAA.)

AIR

I remember it coming as something of a shock to learn as an infant that all around us is not just a kind of emptiness but something physical called air. The thought of air as a thing, tangible and weighty, seemed to contradict everything I'd come to experience up to that point. You can't touch it, feel it or see it; how can it be a thing? As we grow up and piece things together, gradually we work out there must be something physically there: after all, there's wind – a movement of air – and on a blustery day, it's certainly weighty and only too tangible as it buffets us and knocks over trees. Later, we become aware of just how substantial air is when we see how trains and aeroplanes have to be streamlined to overcome the resistance it offers to fast moving things. Although it's invisible, we realise that we are, in effect, ploughing through air all day long, much as boats do through water.

At some point in our education, we are told a little more about this invisible presence: that air is not in fact a single kind of stuff but a mixture of several different things: nitrogen and oxygen mainly with a bit of carbon dioxide (increasing every year, unfortunately). We learn that these invisible things are all called gases and there, for the majority of people who don't continue with science, it largely rests. Gases seem pretty mysterious and, judging by the smell of some of them, pretty unpleasant too.

To get a more sophisticated idea of what a gas is, it helps to try imagining what it is made of: vast numbers of molecules, very tiny units of matter, much too small to see. In contrast to solids and liquids, the molecules in gases are all far apart from each other. They are also not sitting still; they are rushing around freely at enormous speeds, roughly as fast as an aeroplane, bumping into one another or other objects in their path.

Figure 5.3 shows a mixture of two kinds of molecule in a gas (red and blue). It's not to scale – in reality, the molecules are much, much smaller and further apart.

You can find the animated version (https://commons.wikimedia. org/wiki/File:Test_Translational_motion_gif.ogv#/media/File: Translational_motion.gif) in the Support Materials on the book's Routledge webpage, showing how the molecules move around.

The word 'mixture' is used in science to signify that the differing molecules, whether of nitrogen or oxygen or carbon dioxide, are interspersed randomly amongst one another, none attached to another. They move around, spacing themselves out evenly, on average, throughout the space they occupy.

What is less easy to imagine is that there is also an additional kind of molecule present in air: the familiar H_2O molecule: water. Instinctively, we think of water as a liquid, but, pressed to think it through, we know that it can also exist in other forms: as a solid in ice or as a vapour, in humid air. We know that water in a saucer left in the open soon disappears, especially on a

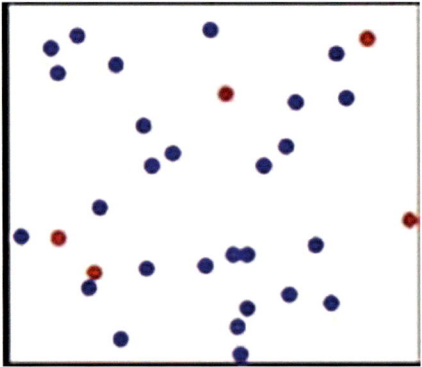

Figure 5.3 Two kinds of molecule in a gas mixture.
(Image credit: A. Greg via Wikimedia (public domain).)

hot day – it invisibly enters the atmosphere. More precisely, H_2O molecules are continuously vibrating and moving around inside liquid water, and the fastest moving ones gradually fly off from the surface of the liquid. There they mingle with those of nitrogen and oxygen already in the air. We call this process evaporation – i.e. becoming vapour. As you might imagine, this process of evaporation is happening all the time at water surfaces – molecules are leaving the top of a cup of tea, a bath, a reservoir or an ocean. The air is constantly filling up with H_2O molecules. But it's also losing them at the same time, as some molecules re-enter the liquid state from the air – a process we know as condensation, the opposite of evaporation (see Figure 5.4).

You're well aware of this when you boil and braise in the kitchen: you soon see the windows misting up with liquid water droplets. You've been busy evaporating H_2O molecules – creating water vapour – while cooking and they've condensed back into the liquid state again on the window. The same can occur in the confined space of a car: molecules of water vapour in your breath get absorbed into the air then condense out as a mist of liquid droplets on the windows. Why does this happen, you might well ask.

You'll have noticed that this 'steaming up' effect seems to occur more frequently when the windows are cold. It's frosty outside, you start the car, the heaters haven't yet done their work and soon your view gets blocked by water droplets condensing on the windscreen. Something similar happens when, out strolling on a bright frosty morning, your breath becomes visible as a cloud of tiny water droplets as it hits the cold air. Steam from a boiling kettle is another everyday example. When

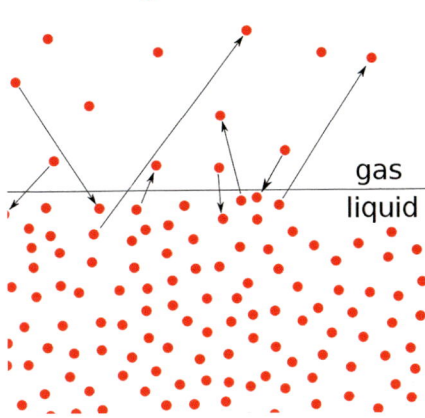

Figure 5.4 Molecules moving between liquid and vapour.
(Image credit: HellTchi via Wikimedia, https://creativecommons.org/licenses/by-sa/3.0/deed.en)

looked at closely, these various kinds of mistiness – steam, breath or fog – consist of a mass of tiny water droplets, suspended in the air. Each droplet is liquid, made visible by light reflecting off it, formed from the invisible vapour as it hits cooler air. The steam puffing out from the spout of a kettle shows us this process happening. If you look closely at it, you'll see a small space close to the spout which is free of steamy clouds. This is the invisible water vapour leaving the boiling hot kettle just before it hits the cool air and condenses into visible droplets.

So, there we have it, water vapour, present everywhere in the air, is in fact an invisible gas. It's just another one of the mixture of gases in air, alongside nitrogen, oxygen and carbon dioxide. Talking about all this in a discussion group one day, it didn't take long for an obvious question to emerge, put succinctly by Mary: 'Why do we call it a vapour then if it's just a gas? Is there any difference between the two?'

VAPOURS AND GASES

In common parlance, we use the two words interchangeably, with vapour as perhaps the more poetic of the two. Shakespeare talks of 'this most excellent canopy, the air….this majestical roof, fretted with golden fire, why it appears no other thing to me than a foul and pestilent congregation of vapours'. He doesn't say 'gases' and that's no surprise: the word wasn't invented till the 1650s, some time after he had died.

Scientists have come to exploit this fortunate duplication of words, to distinguish between two distinct conditions. During the nineteenth century, it was discovered that if you compress a gas hard enough you can, under certain conditions, force it to turn into a liquid – to liquify it. Today, with cylinders of oxygen being delivered to hospitals for breathless patients and butane to country cottages for cooking, we have become familiar with the idea of 'gases' being stored in liquid form, often within strong cylinders that can withstand the pressure. Natural gas is now liquified routinely so that it may be transported in compact form around the globe in ships, rather than as gas through pipelines.

It was soon found, however, that gases can only be liquified under specific conditions: they have to be below a certain temperature and above a certain pressure. Oxygen, for example, must be cooled below −119°C and pressurised to fifty times normal atmospheric pressure for this to occur. That's why we don't ordinarily see oxygen in its liquid state: the everyday world is too warm for it. The same is true for many familiar gases, including nitrogen and hydrogen. The so-called 'critical temperature' is very low indeed – hundreds of degrees below zero Celsius. Exposed to everyday temperatures, liquid oxygen, nitrogen or hydrogen would immediately evaporate. The great exception is water vapour which has a much higher critical temperature than those of most other familiar gases. As we know from watching it condense on windows, water vapour will liquify at everyday temperatures: anything below 100°C and above 0°C under normal conditions.

So, in everyday circumstances, H_2O can be in the liquid or vapour state, whereas oxygen, nitrogen and carbon dioxide must remain gases because, at everyday temperatures – say 20°C – they are way above their critical temperature of −119°C. This leads to the simple distinction: gases are called vapours when they are at a temperature and pressure that enables them to be liquified. For H_2O, this is the case here on Earth where temperatures in temperate regions are typically around 20°C–40°C and pressures around one Atmosphere (or 'Bar'). Our atmosphere can contain water either in liquid droplets or in vapour form, but nitrogen, hydrogen and oxygen only as gases. As an interesting aside for people who follow the news about other worlds, conditions can be quite different further out in the solar system. The atmosphere of a moon of Saturn called Titan is so cold (−179°C) that methane exists as a liquid. It collects in large lakes that have been observed by a space probe.

THE WATER CYCLE

The physics theory we have just explored may seem rather abstract and complicated at first sight: the behaviour of various gases at different pressures and temperatures. It provides, however, the key to understanding important aspects of our weather patterns and helps us see why apparently small changes in the climate can cause such havoc in various parts of the world. The ability of H_2O molecules to cluster together as a liquid or space themselves apart as a vapour underpins the alternating patterns of wet and dry weather common in temperate zones. It enables water to cycle through its phases, first in the oceans, then in the atmosphere, and finally, in the streams and rivers through which it returns. All life depends on this marvellous roundabout.

Before getting back to the story of the 'Pineapple Express' in California, we need to complete our understanding by explaining why rain forms at all, why water vapour doesn't just stay locked up invisibly in the atmosphere.

As the water cycle diagram in Figure 5.5 shows, and residents of mountainous regions know only too well, rain tends to fall mainly in the

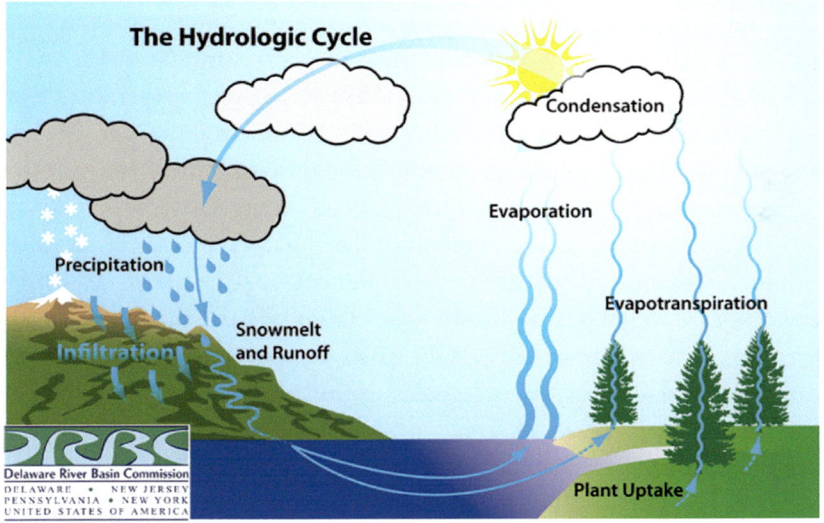

Figure 5.5 The water (hydrological) cycle.
(Image credit: Delaware River Basin Commission, https://www.drbc.gov)

uplands. As winds blow across an ocean, the moving air picks up water molecules evaporating from the ocean's surface. This moist air travels happily across the oceans and flatlands, carrying its load of water vapour without losing any. When it reaches hill country, such as the western coasts of Ireland and Scotland in the UK or the Sierra Nevada in the USA, however, conditions change. Here, the air is forced by the rising land to move upwards. As anyone who has climbed a mountain will testify, the temperature drops significantly as you rise higher.

What happens next is explained by the final piece of theory: the effect of cooler temperatures on moist air. The molecules of which so many substances, including water, are made are not entirely independent entities, unaffected by their neighbours. A moment's reflection confirms this: things hold together – water, wood, plastic, flour – the molecules in substances stick to one another, things don't fly apart. There must be a degree of attraction between molecules just to hold them together enough to form solid (or liquid) things. These attractive forces are relatively slight, as we know because solid things can be broken or cut up and liquids can break up into ever smaller droplets. In a gas or vapour at everyday temperatures, on the other hand, conditions are different: molecules are far apart and moving extremely fast past each other – comparable to the speed of an aeroplane. As a result, the molecules are unaffected by these forces of attraction as they fly past each other.

When molecules slow down and get closer to one another, however, they spend more time in each other's vicinity, allowing them to experience this slight force of attraction long enough to pull them into contact with one another. The molecules are then able to aggregate into what are in effect tiny regions of adjacent molecules: liquid regions. This is what underlies the process we know as condensation. The speed at which molecules are moving in a gas or vapour (more strictly the *average* speed) corresponds to what we experience as its temperature. The higher the temperature, the faster the molecules are moving; the cooler it is, the more slowly they move. As a vapour cools, the molecules of which it is composed move more and more slowly, thus becoming increasingly liable to aggregate into liquid droplets.

Of particular interest to our story is what happens as air moves away from a warmer to a cooler zone: clearly, the molecules of which it is composed must slow down. This is exactly what happens to molecules in air as they are forced to rise up when encountering a mountain range. The temperature falls and the molecules gradually lose speed as the air

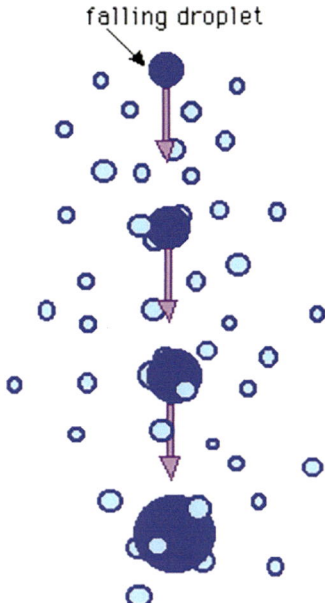

falling droplet

Figure 5.6 Formation of raindrops.
(Image credit: unknown.)

rises. If the mountain is high enough, at some point they will have slowed down enough for the forces between them to draw them together into clusters, tiny regions of liquid water.

These gradually grow into tiny droplets, so tiny that they remain suspended in the rising air, ultimately forming the regions we see as clouds. Under the right conditions, these droplets grow larger and larger until they are sufficiently heavy for gravity to make them fall as raindrops (Figure 5.6).

If the temperature is especially low up there, the water in these droplets may freeze forming crystals of ice (Figure 5.7). This is the origin of the solid kind of precipitation we know as hail, snow or sleet – the magical consequence of water vapour in the atmosphere reaching freezing cold conditions high up in the atmosphere.

H_2O molecules are amazingly versatile. Not only are they able to shift from their spaced-out arrangement in the vapour state to close proximity in the liquid state but are also able to crystallise as a solid when liquid water freezes into ice at 0°C. Once again, it's the force of attraction between the molecules that does the trick, locking the freely moving H_2O molecules of a liquid into the rigid array of a crystalline solid when they are moving sufficiently slowly.

Figure 5.7 Ice crystal.
(Image courtesy of NASA.)

Returning to the plight of Californians (and plenty others around the world), beset with alternating floods, droughts and consequent fires, the so-called 'atmospheric river' represents a serious threat, one that is set to loom larger as the climate emergency deepens. It is a particular instance of a general weather pattern: air, soaked in water vapour drawn from the oceans in warmer climes, hitting a cooler zone and discharging its load. The ability of the air mixture to welcome water vapour molecules where it's warm but to lose them where it's cooler is the very essence of the Earth's beneficent cycle – transferring and desalinating ocean water which then irrigates the parched land. Maintaining this life-support system is critically dependent on the precise balance of temperatures in each region – a balance which humankind is sadly upsetting. As ocean temperatures rise, even by a small amount, evaporation speeds up, loading the air with greater quantities of water vapour. Communicating to the public at large how such an apparently small thing as a degree or two rise in temperature can lead to drastic consequences for our entire way of life is no easy task. Let us hope that leaders across the world are persuaded of the delicacy of this balance and find ways to re-establish it.

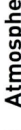

Frost in May – still not safe to put your seedlings out! An unusually cold Spring was taxing the green-fingered members of one science discussion group. Julie had just moved into a city apartment with a garden and was full of horticultural questions. How do perennials survive, she asked – do they have some kind of antifreeze? What appeared at first to be a joke turned out to be not far from the truth. Molecules of the sugar sucrose or of proteins in plant tissue do act as an antifreeze, lowering the temperature at which water in the tissue freezes in perennials, much as salt does when sprinkled on ice.

Picking up on the wintry theme, Patrick wondered how annual plants ever get to succeed in the long run if they fail to make it through the colder months. Why hadn't they died out long ago, he asked? Even as he spoke, he realised the obvious answer to his question: of course, they sow their seeds once they are mature and start off a new generation. And that's where the sex comes in.

REPRODUCTION

Many perennial plants have the huge advantage that they can reproduce themselves vegetatively, without recourse to sex, by developing bulbs or rhizomes or other means. Bulbs, such as those found in tulips and daffodils, are made of modified leaves – familiar as the rings you see in onions. A key function of the bulb is to store food, including the sucrose that prevents them freezing in cold weather. It's energy from this stored food that helps them get through the winter. The bulb also contains a miniature bud inside, just waiting to begin growing (Figure 6.1).

A rhizome differs from a bulb in that it is essentially modified stem tissue, rather than leaf. It runs horizontally underground. New shoots can grow upwards from it, and roots downwards. Like a bulb, it stores nutrients that see it through the winter when the upper parts of the plant die off. Ginger is an example of a rhizome (Figure 6.2).

DOI: 10.1201/9781003272779-7

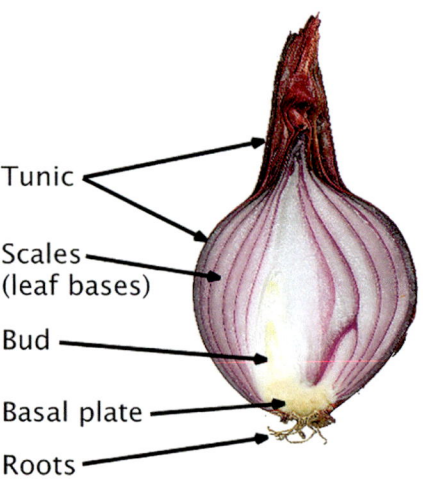

Tunic

Scales
(leaf bases)

Bud

Basal plate

Roots

Figure 6.1 The parts of a bulb.
(Image credit: Amada44 and Peter Coxhead via Wikimedia, https://creativecommons.org/licenses/by-sa/3.0/deed.en)

Figure 6.2 Rhizomes of ginger.
(Image credit: Sengai Podhuvan via Wikimedia, https://creativecommons.org/licenses/by-sa/3.0/deed.en)

A tuber is simply a section of a rhizome that has thickened to store extra nutrients, particularly starch. The latter is a large carbohydrate molecule that holds plentiful energy between the atoms of which it is composed. A model of this can be seen in Figure 10.2 in Chapter 10. Potatoes are a familiar example of a tuber.

Storing food for the winter months and preventing their watery parts from freezing up are not the only tricks perennials have. They also share the important botanical ability to increase their numbers through cloning rather than sexual reproduction. They are able to develop and grow leaves, stems, flowers and roots from unspecialised types of cells (stem cells) known as 'meristem' cells. This particular type of cell simply divides into two to produce daughter cells that are identical to one another. The genes in each daughter cell are the same as those in the parent cell. These daughters, and their daughters in turn, are capable of developing into any of the various kinds of specialised cells that go to make up the tissues of a mature plant.

WHAT'S THE POINT OF SEX?

So, if bulbs and rhizomes get along fine by just growing new cells from old – the new-born springing vegetatively from parental tissue – why don't we all do it: humans, other animals and annual plants? Why do we have to go through such complicated mating processes to bring on the next generation?

Some species reproduce with sex, some without and others are flexible, capable of either. As both methods have survived the rigours of evolution, there must be pros and cons to each. The arguments against sexual reproduction are pretty straightforward: finding a mate can be time consuming, but reproduction cannot occur without one; the process can consume precious energy and the complexities of fertilisation mean the process isn't always successful. So clearly there must be some strong points in favour of sexual reproduction for it to have taken off in the big way it has.

The overwhelming advantage flows from the variation it introduces into the set of genes (the genome), and hence the offspring. Diversity within genes, and of the physiology and anatomy they go on to shape in the individual, can be very advantageous. Over-reliance on one variety of potato in Ireland in the 1840s led to famine when a particular micro-organism was able to blight the entire crop. Any change in the immediate surroundings – food sources, predators, microclimate, disease carriers – that proved deleterious for one individual would immediately threaten the whole population. Genetic variation, on the other hand, means that each individual offspring is slightly different and different individuals may have different propensities to survive. In the long run, survival is not just good for the individual, but also fuels the evolutionary process by favouring the propagation of those characteristics best suited to their environment.

This evolutionary process is illustrated by the example of a light-coloured species of moth. During the period of widespread industrialisation in the nineteenth century in England, the tree trunks on which they habitually settled began to darken. As a result, they became increasingly easy for predators to spot. By chance, some individual moths experienced a random mutation in a gene that made them darker. With their chances of survival dramatically improved, these darker variants lived longer and so produced more offspring. Their descendants ultimately came to dominate.

SEXUAL REPRODUCTION

Plants and animals have differing ways of handling sexual reproduction, but some fundamental elements are common to both kingdoms. The key concept is that characteristics from each of the two parents get mixed randomly as they are passed on to the offspring. This mixing occurs, of course, in every generation: each parent's sex cells – sperm and egg – were themselves created from a random mix of genes from their parents; and so on back through the ancestors.

But what exactly do we mean by the rather vague word 'characteristics'? Our bodies – and those of other animals and plants – develop their shape and functioning thanks to the work of several types of molecule: notably proteins, fats and carbohydrates. Proteins are made within the cells of organisms from the material they eat using the recipes encoded in their genes.

As described in Chapter 4, a gene is a short stretch of the very long DNA molecules that lie in the heart of almost every cell of our bodies – inside the nucleus. Each gene contains the recipe for making a specific type of protein. Proteins, produced in incalculable numbers inside cells, go on to make the tissues of plants and animals and the enzymes that create and regulate them.

The act of reproduction is the means by which genes from one generation get transmitted to the next. The selection of genes that get transferred determine the form and functioning of the new generation. Thanks to sexual reproduction, in contrast to the asexual cloning or vegetative forms, we each have two parents. Each parent supplies just one version of each gene from the two they have and it's random which version gets transferred, introducing variation. Further variation follows as the genes, packaged up in the larger structure of chromosomes, get mixed together randomly when the sperm and egg cells fuse (see Chapter 4). Differences in the genetic make-up of offspring and parent is what has enabled species to evolve and adapt to widely differing conditions across the Earth. This is the pay-off for all the costs of sexual reproduction.

HOW PLANTS DO IT

School biology lessons in the 1950s were hardly riveting in my personal experience. There was one major theme we wanted to know about as young teenagers, and pistils, pollen and stamens were not it. How misguided we were! Reproduction in plants is an amazing story in itself which now, I realise later in life, lies behind the fascination gardeners and nature lovers have with flowers, seeds and propagation. To begin with, it's so varied.

Firstly, there's the option to dismiss sex altogether as plants can choose to do when they rely on bulbs or rhizomes. As described above, meristem cells simply divide to produce the specialised daughter cells needed for making buds or leaves, roots or shoots. The process of specialisation is triggered by hormones and can occur in response to gradual changes in the average ambient temperature or seasonal variation in the length of the day. Plants like potatoes, strawberries or onions can grow in this vegetative way but can also go on to flower and reproduce sexually.

Many other types of plant don't have the vegetative option: sexual reproduction is their only route. They develop male and female parts but in contrast to us humans (and most other animal species), these distinct parts are both found in the same individual, in the flower to be precise. This means that flowering plants don't necessarily have to engage with another plant to reproduce. They have the option to fertilise themselves by transferring pollen, containing male sex cells, from the male part (stamen) directly to the female part (pistil) (see Figure 6.3).

Self-pollination like this does indeed happen but many species have evolved to minimise the chances of it by physically separating the male and female parts (as in the cucumber) or arranging for them to mature at different times or be incompatible in shape or size (primroses, for example). The reason for this is to maximise genetic variation by reducing the chances of genes from male and female parents coming from the same individual plant. With less variation in the gene pool, the risk is greater that all the offspring will be susceptible to the same disease.

Cross-pollination ensures that reproduction involves different parents: genes from the pollen (male) in one plant combine with genes in the

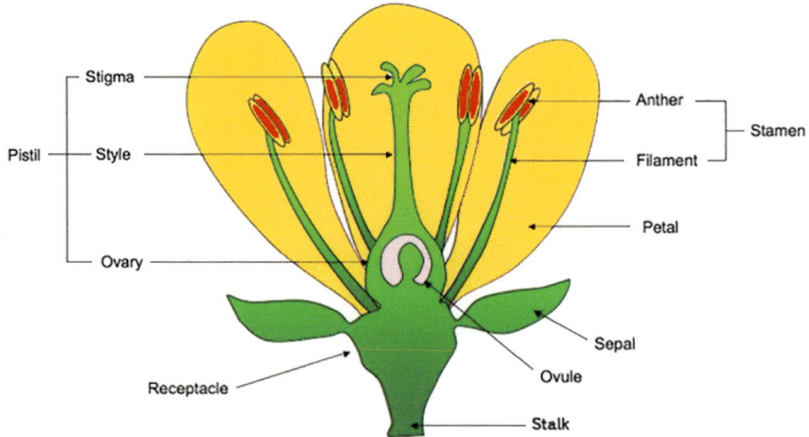

Figure 6.3 The parts of a flower.
(Image credit: Anjubaba via Wikimedia, https://
creativecommons.org/licenses/by-sa/4.0/deed.en)

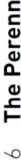

eggs (female) of another. As a result, a greater range of versions of each gene are mixed in producing offspring. This increases the chance of producing some progeny that are particularly fit. But how does the mixing take place at all, given that plants are largely immobile and don't get out to meet their partners? Evolution has resulted in many ingenious ways of overcoming this mobility problem.

One involves a helpful force of nature – the wind (where it prevails). For many species of grass and tree, lightweight pollen on the outer surface of flowers gets caught by the wind, blown about and, with luck, falls on the female parts of an individual of the same species, elsewhere. Such wind-borne fertilisation requires no flowery means like bright colours or a seductive scent to attract a pollinator. The wind either blows or it doesn't, so pollen production sometimes occurs before the leaves grow, which would otherwise impede the passing air currents.

The most obvious vectors for moving pollen around are of course animals. They are not only mobile but, unlike the wind, are open to persuasion. A sweet-smelling nectar and a brightly coloured flower are all it takes to attract bees, butterflies, bats and birds who through their peregrinations can connect distant reproductive partners. The interplay of winged beasties with the sex cells of plants is an inspiring example of collaboration arrived at through evolution. Flowering plants produce energy-rich nectar that fuels the pollinator, while hairy parts of the legs

Figure 6.4 Bee picking up pollen.
(Image credit: Lars Falkdalen Lindahl, https://
creativecommons.org/licenses/by-sa/3.0/deed.sv)

and underbellies of the creatures grip the pollen grains which convey the
vital genes from the male part of the flower (Figure 6.4).

Co-evolution has not only brought this mutually advantageous system
about but has gone on to develop a close match between the size and
shape of the mouthparts of the pollinator and the corresponding parts
of the flower. This not only allows for more efficient transfer of pollen
but also prevents other freeloader species from raiding the source of the
nectar. This precise matching of flower shape and pollinator mouth may
be one of the factors driving the evolution of such variety in flowering
plants. Other species of plants are pollinated by moths; their flowers
tend to have a flat shape which allows the insects to land comfortably.
Others, such as certain orchids, which rely on birds like hummingbirds
for pollination, have flowers shaped in such a way as to enable the bird
to approach the flower without catching their wings in it.

It's not only the form (morphology) of flowers that has evolved in
partnership with specific animal species, their colour has too. Bees, for
example, are unable to see the colour red, so flowers that attract them
are likely to be blue or yellow in colour. Flowers that emerge at night,
mainly in hotter countries, may use bats for pollination; their flowers
tend to be large and white to stand out in the gloom.

Some flowers go so far as to attract a male insect by mimicking the
appearance and smell of the female of the species. The bee orchid, for
example, has a petal that looks like a particular kind of female bee

Figure 6.5 Bee orchid.
(Image credit: Bernard Dupont via Wikimedia, https://creativecommons.org/licenses/by-sa/2.0/deed.en)

(Figure 6.5). Male bees attempt to mate with it and inadvertently pick up pollen which they then transport to the scene of their next 'conquest'.

OFFSPRING

Pollination, the task of physically transferring genes from the male parts of one flower to the female parts of another, is a prelude to the all-important process of fertilisation. Once the pollen grains of one flower have been artfully deposited on the female parts of a neighbouring one the process begins. The arrival of the pollen grains causes a tube to develop within the female part of the flower (ovary) through which sperm cells pass to get to where the eggs are (the ovule). Once inside this area, the sperm and egg cells fuse, mixing together genes from the two parent plants. The fertilised egg cell then develops into a seed, which provides a wonderful nurturing environment for the growing embryo. It has an outer coat to protect the embryo and a store of food upon which it will feed initially (Figure 6.6). The ovary itself, which contained the ovules, develops to become the fruit. The job of the seed is to nourish the embryo initially, while that of the fruit is to protect the seeds and help disperse them to a suitable environment.

Dispersal again reflects the marvellous possibilities of co-evolution. Some fruits are light and have wings or hairs that catch the wind. Others are buoyant and float away; many have attractive fleshy parts that get

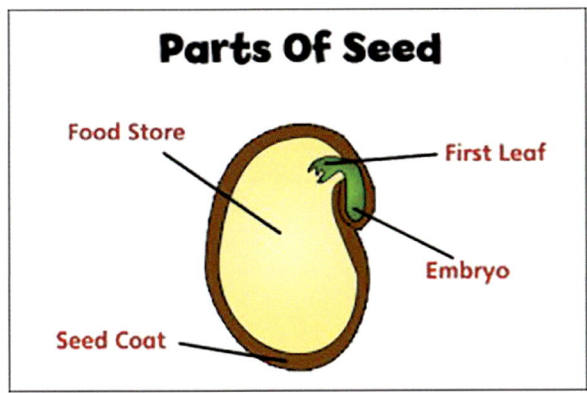

Figure 6.6 The parts of a seed.
(Image credit: H. Guthrie Chamberlain III.)

eaten, enabling the seeds they contain to be excreted some way away; and some are simply sticky, allowing them to attach to any fur that passes by.

Wherever they end up, seeds lie dormant waiting until conditions in their new environment are favourable. During the dormant period, life inside the seed grinds to a halt. It awaits the arrival of water, oxygen, light and warmth. Water will soften the seed's outer coat, enabling the embryo inside to swell and split open the coat. Oxygen absorbed from tiny spaces within the surrounding soil enters, enabling respiration to begin. In this chemical process, the absorbed oxygen reacts with carbohydrate molecules in the seed's internal food store, releasing energy in the process. This enables the seed to germinate. Different varieties of seeds respond to temperature and light in different ways. Some require warmth for germination, others need cooler conditions; some require a certain level of light, others do not. Absence of light can cause a seed to wait patiently for years in the dormant state – a necessary precaution, as any seedling that developed would require light in order to grow.

CONCLUSION

This story of reproduction and its intricacies tells us something of the way in which life is perpetuated: by bringing on the next generation and, in the very long term, by evolving new ways. The form and functioning of every individual, from a single-celled amoeba to you and I is determined (largely) by the information coded in our genes. Duplicating this in a clone enables individuals to reproduce relatively quickly but can spell disaster for a population, if and when its environment changes.

Sexual reproduction has evolved as a result of better survival and reproduction rates when different versions of genes are available from each of two parents. But the persistence of both sexual and asexual reproduction in plants, fungi, sponges, bacteria and a range of other species shows that the balance of advantage is a fine one. For *Homo sapiens* and most animal species, however, the only way is sex.

Seven

The crisp and silvery scene in Figure 7.1 was captured on a garden pond in Worcestershire one winter morning. The crunchy reeds and diffuse reflections appealed to the senses, but it was the strict linearity of the markings in the ice that aroused my curiosity. It seemed at odds with the disordered weeds and crumpled reeds nearby.

The extraordinary truth is that this geometry, at an everyday scale, reflects the order that exists amongst atoms, on a scale a billion times smaller. Crystalline regularity connects structure at the nano-level with features on a human scale. Molecules and atoms of many kinds will aggregate in regular geometric arrays – most metals and minerals, as well as water (H_2O) molecules, are examples (Figure 7.2).

A single molecule of water has a very simple structure. It comprises two small atoms of hydrogen (white) bonded to a slightly larger atom of oxygen (red), as the model in Figure 7.3 indicates.

The bonds that hold together the three atoms of each molecule are strong. That's why water normally remains as water, rather than collapsing into hydrogen and oxygen, from minute to minute. Any molecule is a tightly bound assembly of atoms.

A drop of water or piece of ice consists of zillions of H_2O molecules, lying cheek-by-jowl. It's the fact that H_2O molecules don't ordinarily break apart that enables us to deduce that the three atoms within them must be tightly bonded to each other. But does the same apply to the liquid as a whole? Given that we know pools of water and blocks of ice hold themselves together in ordinary circumstances, we can also deduce that each molecule must be somehow connected to its neighbours. If the molecules didn't stick together in this way, water (or any other liquid) simply wouldn't cohere as a substance at all – each separate molecule would simply drift away.

These bonds that link together adjacent molecules are, however, of a much weaker kind than the ones that hold together the atoms making up each molecule. They suffice to hold together molecules in a solid or

DOI: 10.1201/9781003272779-8

Figure 7.1 Early morning ice on a garden pond.

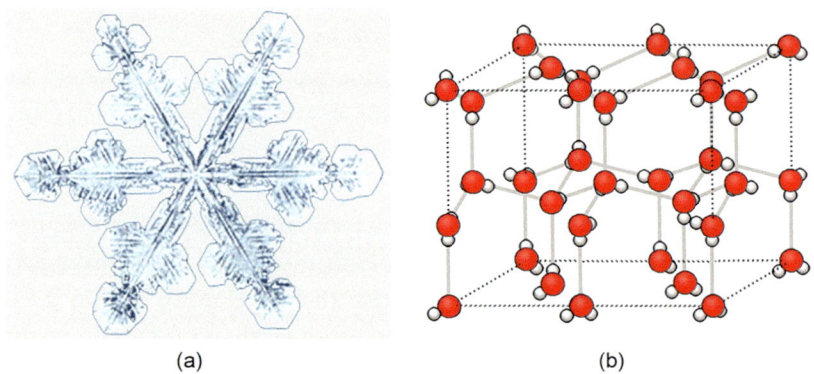

(a)   (b)

Figure 7.2 Crystalline regularity at the molecular level.
(a) Snow flake. (Image credit: Wikimedia (Public Domain).)
(b) H_2O molecules in ice.
(Image credit: IgniX via Wikimedia, https://creativecommons.
org/licenses/by-sa/3.0/deed.en)

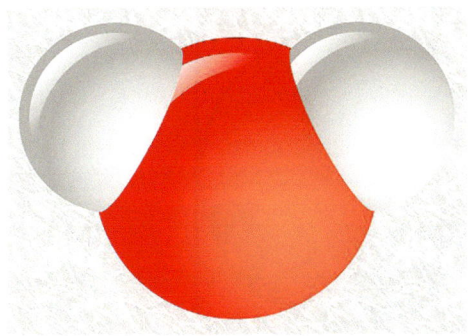

Figure 7.3 Model of a molecule of water, H_2O.
(Image credit: Sakurambo via Wikimedia (Public Domain).)

liquid state much of the time but, as we know from everyday experience, these bonds can be overpowered, under certain conditions. For example, when liquid water is exposed to the air, the molecules near the surface can simply drift off into the air – that's evaporation. Molecules at the surface can separate from each other and disperse amongst the widely spaced molecules in the air. In another example, we know that solid ice can melt, if the temperature is above zero degrees Celsius. In this case, the bonds that hold the H_2O molecules tightly together in the lattice pattern of solid ice are overpowered. This enables the molecules to free up sufficiently to move past each other, whilst remaining close to one another, like a collection of marbles in a pocket. This is the nature of the liquid state, in contrast to the solid. Figure 7.4 illustrates the way molecules are disposed in the three states of matter.

Whether in the solid, liquid or gaseous state, molecules are always on the move in one way or another. In gases – air for example – they buzz around randomly at high speed. In liquids, they slide past each other; in solids, they vibrate and stretch in various directions. These kinds of motion are demonstrated helpfully in three online animations, (links below), in which the movement has been slowed down dramatically to illustrate the point. In reality, molecules in air, for example, are moving faster than most aeroplanes, roughly a thousand miles per hour.

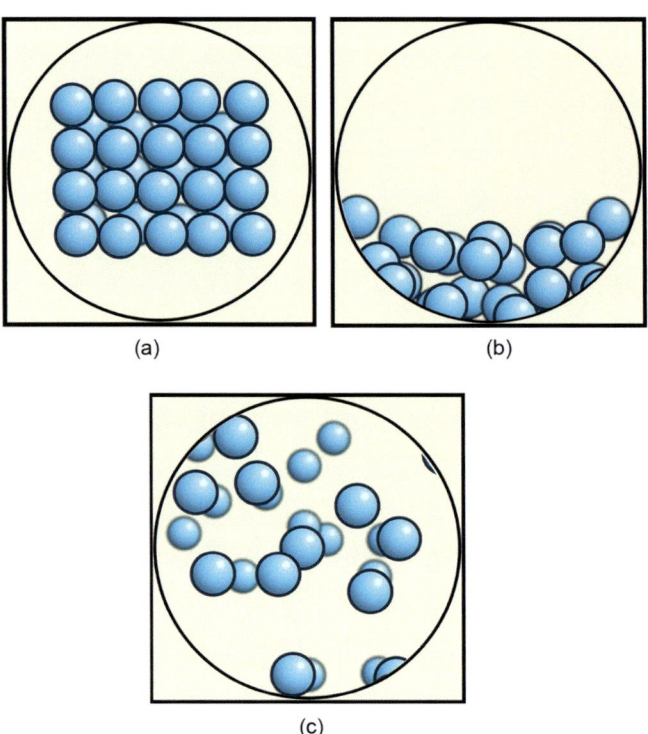

Figure 7.4 Molecules in (a) solid, (b) liquid and (c) gas state.
(Image credit: Students at Learning Tec de Monterrey,
Mexico City.)

Animated versions available at:

Solid (https://upload.wikimedia.org/wikipedia/commons/b/ba/Gif_
AtomosSolido_01.gif)
Liquid (https://upload.wikimedia.org/wikipedia/commons/5/5b/
Gif_-AtomosLiquid_03.gif)
Gas (https://upload.wikimedia.org/wikipedia/commons/2/29/Gif_
AtomosGas_02.gif)

THE SPECIAL CASE OF H₂0

Water is rather a special case amongst liquids. The relatively weak bonds
discussed above, which serve to keep molecules in contact with each

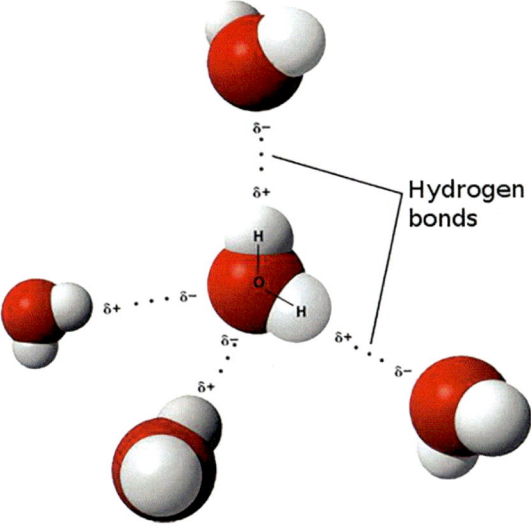

Figure 7.5 H_2O molecules linked together by hydrogen bonds.
(Image credit: Wikimedia, Public Domain.)

other, are reinforced in the special case of water by an extra bond. This special type of bond connects an oxygen atom on one H_2O molecule with a nearby hydrogen atom on a neighbouring one. It's an electrical attraction caused by a small excess of negative and positive charge on each of the atoms (see Figure 7.5).

These special bonds, known as hydrogen bonds, have a middling strength: weaker than the bonds that hold atoms together in molecules but stronger than the forces between one molecule and its neighbours. They are weak enough that, at temperatures above the freezing point, the bonds are overpowered by the buzzing around of the molecules. Above 0°C, water is, as we all know, liquid: a state in which molecules remain close to one another but move freely around each other. As temperatures reduce, however, molecules slow down. At 0°C (32° F), the movement of molecules is sufficiently slow for the hydrogen bonds to bring them to a halt. Each molecule clicks into place alongside its neighbours, with

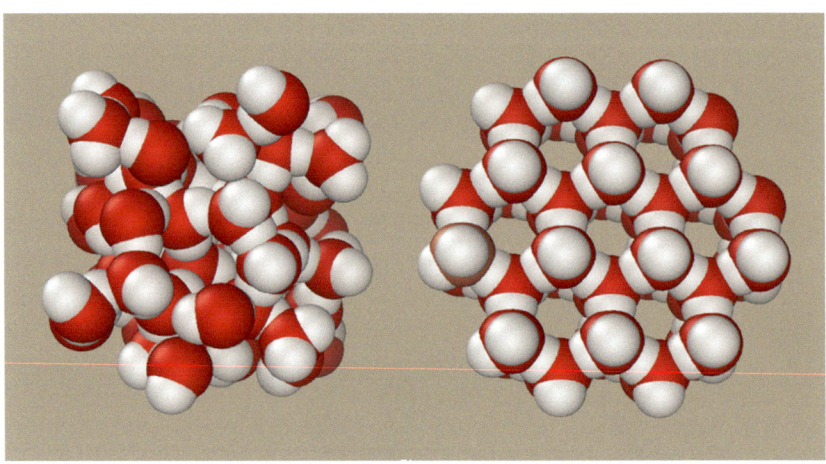

Figure 7.6 H_2O molecules in water (left) and ice (right).
(Image credit: P99am via Wikimedia, https://creativecommons.
org/licenses/by-sa/3.0/deed.en)

the hydrogen bonds forcing them into a regular lattice, like eggs in a tray (see the right-hand side of Figure 7.6).

This special kind of bond – the hydrogen bond – is not only important in the case of water, but it also links together a range of other molecules. In particular, it plays a vital role in the giant molecules upon which life itself depends. Being of intermediate strength, hydrogen bonds are able to add crucial structure to the long, thin chain-like molecules of proteins and DNA. The gentle nature of the bond enables chain-like protein molecules to fold up into compact globular structures, suited to their roles as enzymes and hormones, for example. It also enables the two long, thin strands of a DNA double helix to stick to one another, but also to come apart when necessary, to enable their code to be 'read'.

THE ICY POND

Back to the pond….. it's winter, it's night time, the temperature is falling. Soon the chilled air cools the surface water. The H_2O molecules are slowing down and, as the temperature reaches zero degrees, are poised to click together into the rigid crystalline form of ice, thanks to their hydrogen bonds. Crystallisation doesn't happen instantly, however, as you might expect; some kind of initiator is necessary to kick off the process of rearranging atoms. The temperature drops a little further, just below zero; momentarily, the water becomes 'supercooled'. Like a land-slide awaiting the crucial rockfall, the molecules are poised for action. A particle of sediment or the surface of a bacterium is all that's needed, as a former, to enable one molecule to line itself up with the next, a third to join them, and so on, as a cascade of crystallisation is launched.

As the process takes off, the first patch of ice propagates through the supercooled water at the surface. Blade-like crystals grow rapidly across the surface as ice begins to form. Elsewhere on the pond, a different fragment initiates a second rapid crystallisation, then another … Soon blade-like zones of crystallisation are speeding across the surface, randomly orientated with respect to each other, but each strictly geometric in itself. It was the sheer regularity of the shooting lines on the surface of the pond that caught my attention at the beginning of this story. Now we can see that these markings, on an everyday human scale, reveal the extraordinary regularities that bind together the ultimate particles of which all matter is composed.

BETWEEN THE MOLECULES

The nature of ice was the topic under discussion in a group one day. Sonya had pointed out that a bottle of liquid can burst if put in a freezer, because the ice takes up more space than the liquid. Terry queried whether it would weigh the same after freezing, then realised that it must do, as nothing had been added in the process; the volume must simply be larger, not the mass of water. The two models in Figure 7.6, comparing the way in which H_2O molecules are arranged in liquid water and ice, indicate why this is so. In ice, the molecules are arranged in a crystalline lattice, like eggs in an egg box. The hexagonal arrangement of the crystal means the molecules are held slightly further apart than they would be in liquid water. As a result, they occupy slightly more space, which can indeed cause a bottle of water to burst in a freezer.

'Water is weird' was the reaction from Mary; 'you'd expect stuff to get more dense when you go from liquid to solid'. She was right; scientists describe the behaviour of water as anomalous. As Julie conjectured correctly, most other liquids would contract – get more dense – on freezing. Anomalous though this behaviour may be amongst liquids, it plays a crucial role in maintaining the health of our planet. Much aquatic life, for example, depends on it. Because ice occupies a little more space than the equivalent quantity of water, it is less dense than water. As a result, it floats on top of a pool of water – as ice cubes do in a watery drink.

On ponds and puddles and lakes, this results in ice forming at the surface rather than in the depths, enabling life to continue below, even on a freezing cold day.

The anomalous nature of water is also of crucial importance more broadly, in maintaining our delicately balanced climate. The massive ice fields in the Polar Regions float on the surface of the oceans because they are less dense than the liquid water beneath them. The reflective white surface they present ensures that a proportion of the heat radiating from the Sun is reflected back, helping to maintain moderate temperatures on the planet.

Important though the role of water and ice are in maintaining the health of the planet, it was in quite a different direction that the discussion turned in the group at this point. Unexpectedly, Sonya said: 'it freaks you out that there is just a vacuum between molecules in ice'. 'I suppose it's the same as air being 99% empty space. Is it really a vacuum or just that we don't know what it is?', she continued. By homing in, not on the atoms or molecules of H_2O but on the spaces in between them, the group had stumbled upon a deep and difficult truth: the extreme emptiness of the universe. Matter occupies just a tiny fraction of the entire volume of space. We humans, living on the thin crust of our tiny planet, are in constant contact with matter in its solid, liquid and gaseous forms. In between, the atoms of which it is composed and out beyond our island world, emptiness prevails. Indeed, the very atoms of which matter is composed are almost entirely empty themselves; the particles of which they are made occupy but a tiny fraction of the volume of an atom – as a particle of dust to the Albert Hall, as one writer vividly explained. The rest is a vacuum, empty space.

These realisations about the extreme emptiness inside atoms came as a complete surprise to scientists, as much as lay people, when it became apparent through experiments conducted over a hundred years ago. The expectation had been that atoms would be made of more solid stuff – the name 'plum pudding' was given to a model of the atom popular at the time. This image expressed an idea that the negatively charged particles (electrons) recently discovered inside atoms might be embedded like raisins (known then as plums) in a kind of 'dough' of positive charge. The solidity of a pudding seems to capture the intuitive idea that atoms should be hard and stiff to explain the nature of everyday materials.

Today, we are obliged to accept the uncomfortable truth that solidity isn't quite what it seems. The interior space of atoms is largely empty and the universe too. Material particles occupy just a minute fraction of the volume of otherwise empty space. Despite this, atoms made up of these particles behave as hard things, colliding and bouncing off each other and aggregating to form the solids, liquids and gases of which our world and the universe are made. The early twentieth century, in which so many of these discoveries were made and theories developed, has bequeathed a more unsettling picture of the ultimate nature of our material world than many earlier philosophers and scientists had imagined. The quest to understand it continues apace. Once again, a question about ordinary things – icy ponds and bursting bottles – has led into profound thoughts about the nature of our universe.

Eight

The alluring image in Figure 8.1 caught the attention of a science discussion group recently. Looking at first sight like a rich middle-Eastern tapestry, it is in fact a model of part of a typical human cell put together using information from various instruments used to study tiny things. The image was constructed by an artist based on data drawn from a variety of sources: microscopes, X-rays and NMR (nuclear magnetic resonance) machines.

Its aesthetic appeal drew the group members in and the complexity of its composition provoked questions about what it represented, what the intricate structures were. Conventionally, images of cells in books and websites emphasise the great variety of their parts and their intimidating technical names (Figure 8.2).

Learning the functions of the parts of a cell and memorising their names is an arduous part of biology lessons at school, but, for the discussion group, it was the system as a whole that captured their attention. How do all the bits interact? Are they busily working away all the time? How many cells have we got in our bodies? How big are they?

Advances in animation technology and biological imaging have begun to transform the way we understand cells. Traditionally, they have been represented in a two-dimensional way on the pages of a book, based on actual or imaginary slices through the cell. Such representations are necessarily flat and static. In reality, however, cells are voluminous and flexible three-dimensional containers, simply buzzing with activity. The structures inside them (called organelles) and the molecules that make them up are also lively, three-dimensional entities in a state of regular movement. This activity has now been brought alive for us in glorious 3-D technicolour animations.

An example from Harvard University — *The Inner Life of the Cell* — shows molecular structures interacting inside a typical cell. Although the animation doesn't explain what's happening, it does give a helpful impression of the shape of structures and the way in which they interact with

DOI: 10.1201/9781003272779-9

Figure 8.1 Model of part of a human cell.
(Illustration reproduced courtesy of Cell Signaling Technology, Inc. (www.cellsignal.com).)

one another. One example, available at https://xvivo.com/examples/the-inner-life-of-the-cell/, follows a white blood cell's movement along the lining of a body tissue and its response to an external stimulus.

SCALE

A first step in thinking about cells and their contents is to get a sense of the relative size of different microscopic things. Molecules, cells, viruses, bacteria and DNA – all are very small, undetectable by the naked eye; but which is smaller than which?

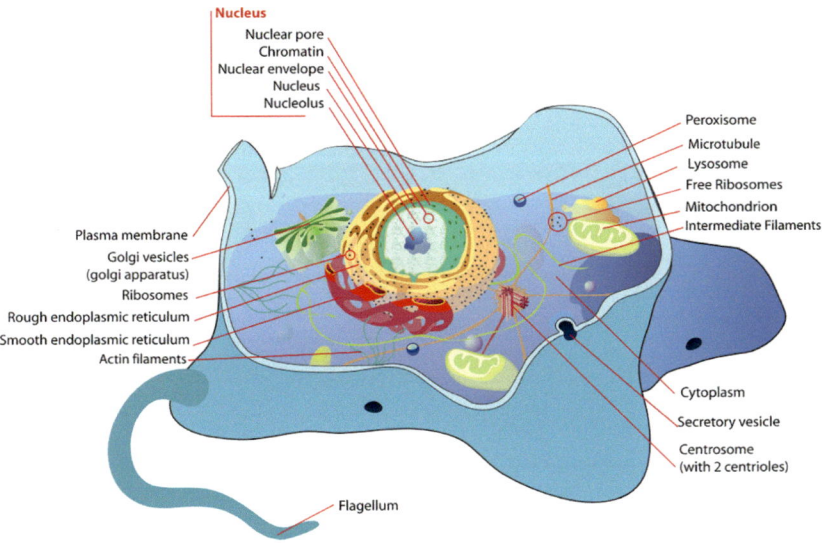

Figure 8.2 The parts of an animal cell.
(Image credit: LadyofHats via Wikimedia, Public Domain.)

Let's start with the cell – it's the main unit from which animals and plants (and other organisms) are built. Individually, cells are created from parent cells that divide, then develop, reproduce themselves and die. By sticking together, they form the tissues of which organs are composed (though there are some exceptions to each of these points). Cells come in numerous types which vary enormously in detail but all share an important common feature: an outer membrane which separates the inside from the outside. The watery world inside a cell and the different watery world outside are kept apart by an oily boundary (cell membrane) that protects the workings of the interior and controls what enters or leaves the cell. What goes on inside is quite unlike what happens outside. The boundary is usually closed to all but carefully selected molecules. Incidentally, it's the ability to break through this protective outer membrane that enables viruses to multiply so prodigiously and wreak havoc within the cells they enter.

Cells are relatively large items at the microscopic level – typically measured in micrometres (thousandths of a millimetre), though sizes do vary considerably. The structures inside a cell are typically ten to a hundred times smaller. Many of these structures are built out of giant molecules, often proteins, which might themselves be ten to a hundred

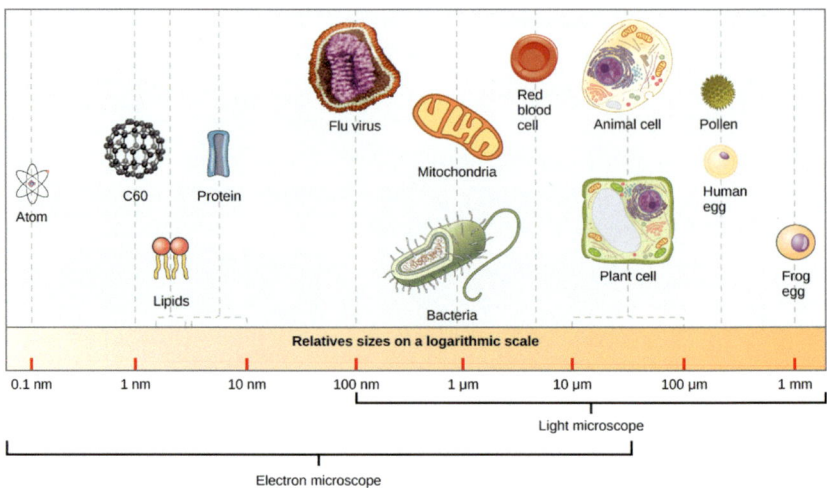

Figure 8.3 Relative size of cells and their components.
(Image credit: Charles Molnar and Jane Gair Concepts of
Biology (https://pressbooks.bccampus.ca/conceptsofbiologymol
narcamosun), https://creativecommons.org/licenses/by/4.0/)

times smaller still. Some living organisms, like bacteria, consist simply
of a single cell. The interior of plant cells and animal cells have much
in common but also differ in important ways. Both have a surrounding
cell membrane and a compartment (the nucleus) for their all-important
DNA, but, as just one example of the many differences, a plant cell has
the capacity to capture energy from sunlight using a component which
animal cells lack. In this chapter, we focus on the animal cell.

Viruses check in somewhere between the size of a cell and a giant
molecule. A coronavirus, for example, is composed of hundreds of giant
molecules and measures about a tenth of a micrometre – perhaps a
hundred times smaller than a typical cell. Figure 8.3 gives an impression of relative sizes. Note that the steps in the horizontal direction are
not equal – each one is ten times larger than the previous one. Size is
measured here in nanometres (nm) and micrometres (μm); for comparison, an average sheet of paper is approximately 100,000 nanometres
or 100 micrometres thick.

TISSUES AND CELLS

Cells aggregate together to form tissues of various types – connective
tissue or muscle tissue, for example, in the case of animals. Organs, such

as kidneys or lungs, are often made of several different kinds of tissue. In the heart, for example, we find muscle tissue, connective tissue and nerve tissue. Cells aggregate to form a tissue and different tissues come together to form an organ.

To survive, a complex organism needs to carry out a huge range of different functions. To serve all of these, cells of many types are required, each adapted for its specialised function. Nerve cells transmit electrical pulses, for example, muscle cells contract and red blood cells carry oxygen around. Different kinds of cell have different shapes and are of different sizes. Figure 8.4 contains images of three kinds of blood cell taken with an electron microscope (a type of microscope with very high magnification) and artificially colourised. It shows a red blood cell (doughnut shape), a platelet (smallest) and a white blood cell (fuzzy ball shape).

The highly magnified images in Figure 8.5 are of muscle cells and a nerve cell (neuron) from the brain.

As these images demonstrate, the shapes of different kinds of cell vary enormously. Their forms may even express something of their function: the voluminous red blood cell contains bags of haemoglobin molecules;

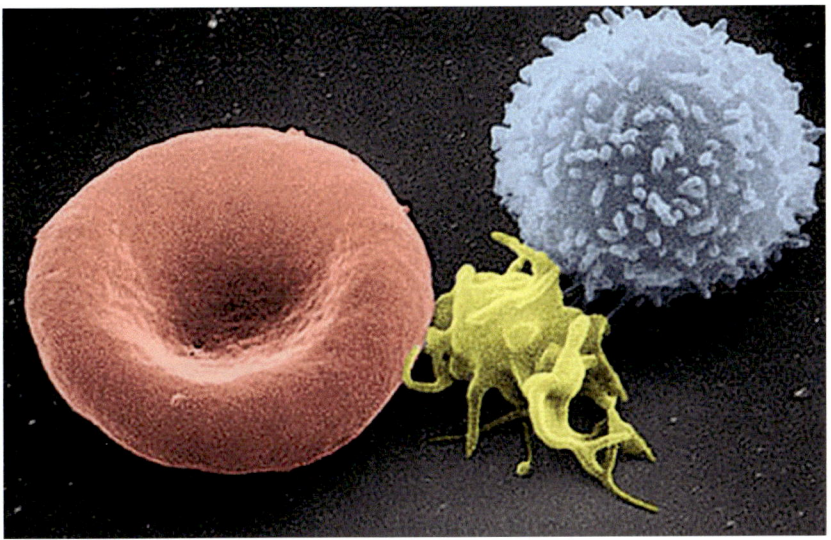

Figure 8.4 Three kinds of cell in the blood.
(Image credit: The National Cancer Institute at Frederick, http:// creativecommons.org/licenses/by-sa/3.0/deed.nl)

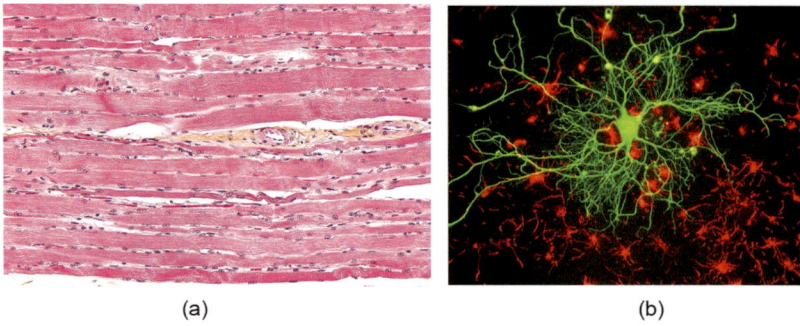

Figure 8.5 Cells have different shapes. (a) Muscle cells. (Image credit: nephron via Wikimedia https:// creativecommons.org/licenses/by-sa/4.0/) (b) Nerve cell. (Image credit: Gerry Shaw https:// creativecommons.org/licenses/by/2.5/deed.en)

the long, thin muscle cell elongates and contracts; the wiry looking neuron transmits electric pulses.

Cells are mostly, of course, tiny – powerful microscopes were needed to create the images above. They vary in size enormously: red blood cells being typically a few micrometres (thousandths of a millimetre) across, whereas some nerve cells are a metre or more in length. The enormity of the number of cells of which we are made is as difficult to grasp as the tiny size of the average cell. It's reckoned that there may be 30–40 trillion in a human being and that's without counting the bacteria we house. That means there are several thousand times more cells in one human body than there are people in the world.

INSIDE THE CELL

Figures 8.1 and 8.2 give an overall impression of the internal components of a cell (known as organelles – small organs). The animation mentioned earlier, *The Inner Life of the Cell* (available at https://xvivo.com/examples/ the-inner-life-of-the-cell/), gives some idea of the activity that goes on inside these structures. Tiny as most cells are, they are complex environments, filled with smaller units, each with a vital role in maintaining the life of the cell.

Essential to life is the cell's ability to take in and use energy. The components inside cells responsible for this are discrete bodies, each with their own outer membrane, called mitochondria. Remarkably, all organisms, not just animals, make use of the same energy-rich fuel: glucose. It's a simple molecule, a kind of sugar, that the body obtains by breaking down the molecules in the food we eat. It travels to all the cells of the body via the blood – that's why a simple blood test is able to reveal the level of glucose in your body, key to diagnosing diabetes (see Chapter 10).

There are many mitochondria in a typical cell (maybe thousands). Inside each is a set of enzymes which make a chemical reaction happen by combining the energy-rich glucose from food with oxygen from the lungs. Rather like burning oil or coal in air, this releases energy, which immediately gets stored in a high-energy molecule common to all forms of life (known as ATP). This ubiquitous molecule takes up the energy extracted from our food and holds onto it, ready for use wherever and whenever it is needed, 24/7.

This is why we need to breathe in oxygen – every cell requires it all the time to release energy, just as a motor vehicle does to burn petrol or a campfire to burn wood. This process, analogous to combustion in the case of fuels, is called respiration and is common to all living things. As a result of the chemical reactions involved in respiration, carbon dioxide and water are produced as waste products (Figure 8.6). That's why we end up expelling these particular substances through our breath and urine.

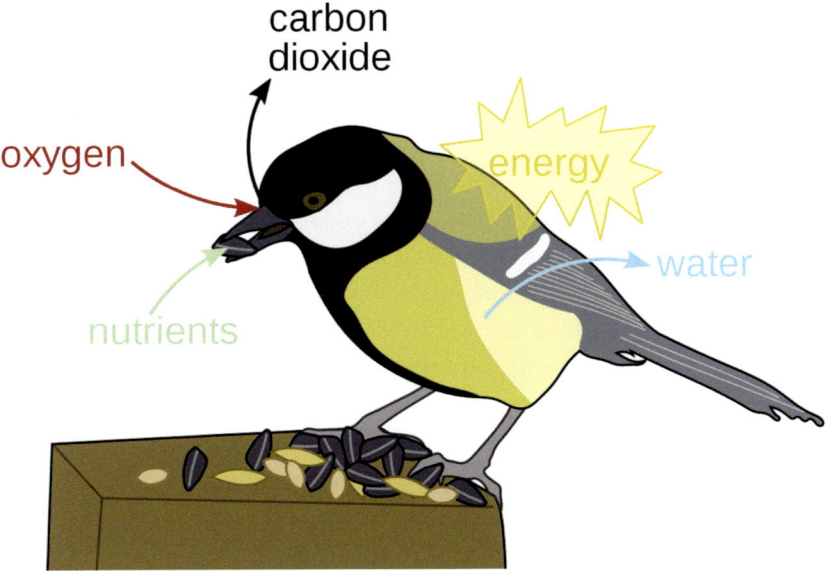

Figure 8.6 Energy through cellular respiration.
(Image credit: Nefronus (Public Domain).)

Another vital function of the cell is to store, and ultimately reproduce, the genetic information that it will need to pass on when it divides into daughter cells. This information resides in the enormously long DNA molecules, held in another organelle: the nucleus. Most cells have just one nucleus. Like mitochondria, it is also separated from the rest of the cell by a membrane of its own.

In everyday life, we get to hear references to DNA in connection with our personal identity and ancestry. That's because the precise structure of our DNA molecules is unique to each of us, hence its use as an alternative to fingerprints for identification. Commonalities in the DNA of particular populations have also proved useful to anthropologists, enabling them to trace the movement of peoples around the world over large spans of time. But DNA molecules are not just waiting around to be used as identity markers or even for the occasional, but vital, moments in which they are transferred from one generation to the next. They are in continuous use, all day and night, providing the information needed to make the proteins required for day-to-day functioning; this is happening rapidly and regularly in our cells.

As mentioned in Chapter 4, the long DNA molecule is divided up into thousands of discrete short stretches known as genes (plus other regions in between). Each gene carries information that specifies the make-up of one particular type of protein. One gene carries the code for a digestive enzyme, another the code for a muscle protein, another for a particular type of antibody, and so on. Altogether, there are tens of thousands of types of protein in the body; information stored in DNA in the nucleus is being used continuously to produce them.

Apart from the mitochondria and the nucleus, another important type of organelle is the place where proteins are actually produced, based on information encoded in the genes. Known as ribosomes, these giant molecular structures float around in the gel-like environment inside a cell (known as protoplasm). A typical cell in our body may contain literally millions of ribosomes. As the model in Figure 8.7 shows, large assemblies of molecules, such as ribosomes, may be made of hundreds of thousands of atoms (the orange and blue balls).

It is clear that the inside of a cell is a complex and busy place! We've introduced just three of the many kinds of types of organelle: the mitochondrion, nucleus and ribosome. What struck members of the discussion group at this point, however, as they grappled with the internal complexity of the cell, was not the desire for further detail, but rather,

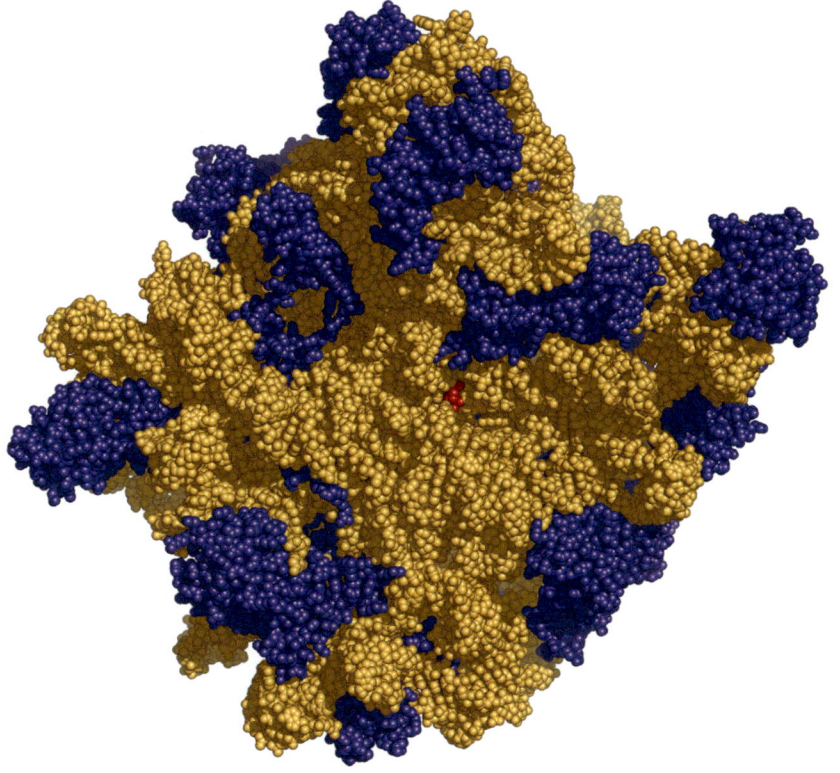

Figure 8.7 Ribosome: a model of one subunit.
(Image via Wikimedia, https://en.wiki pedia.
org/wiki/en:GNU_Lesser_General_Public_License)

the need to pause and reflect on the almost unbelievable facts already presented: the sheer numbers and sizes involved. In our bodies, there are literally trillions of cells, each of which contains millions of 'moving parts'. Powerful microscopes are needed to visualise them. As for the even smaller and more plentiful molecular structures that carry out activity within the cell, even more sophisticated equipment is required: electron microscopes and X-ray diffractometers, for example. It's difficult to take in the miniscule size of the structures that make our bodies work and the enormous number of them at work in each of us.

MOLECULES AT WORK

The multitudinous structures that make up our cells are mostly assemblies of giant molecules, such as proteins and DNA. These structures are not, however, static, like buildings or bridges, but dynamic and flexible, engaged in unceasing activity.

The animation mentioned above (The Inner Life of the Cell (https://xvivo.com/examples/the-inner-life-of-the-cell/)) portrays large protein assemblies moving in and around a cell. It gives a vivid impression of some of the actions of which proteins are capable. In one sequence, a long, thin molecule is shown as having two 'legs' each capable of shifting between two positions. As one 'leg' of the molecule and then the other attaches to a long molecular fibre in alternation, it gives the impression of 'walking' along the fibre.

In a later sequence, one protein (an enzyme) cleaves a long thin molecule into two – breaking bonds between atoms. This is what an important class of enzymes do; it's how the giant molecules in your food get broken down into simpler ones, for example.

Yet another kind of activity is the passage of molecules in and out of cells, through their membranes. Another fascinating animation, https://www.youtube.com/watch?v=lcHy8THENXo, shows how antibodies manage to neutralise a flu virus. The virus enters a cell through its membrane, then discharges its genetic material, RNA (related to DNA), which then passes through another membrane into the cell's inner sanctum, its nucleus. Here, many copies of the viral RNA are made in readiness for a new generation of viruses that will burst out and spread the infection. The animation shows the new virus, while outside the cell, being attacked by smaller antibody molecules which attach themselves to parts of its outer surface. This prevents the virus from entering the cell's nucleus to replicate itself. Unable to replicate and spread to other cells, the virus is effectively neutralised.

WHAT TO MAKE OF IT ALL

The question that arose, as the group contemplated the intricacy of activity within a cell, was just how rapidly it all happens – is it incessant, does a cell ever rest? As Mary put it 'are these tiny molecular activities going on all the time – when I move my hand or when I sit still for example?' Activity inside the cell is indeed happening all the time; the cell never rests. The pace of activity, though, varies. Enzymes, which break bonds and alter the shape of molecules, act as catalysts. Some perform hundreds of reactions in a single second, whereas others only manage one every minute. For comparison, an average typist manages just over three characters in a second – maybe a hundred times slower than the actions of a fast enzyme. On the other hand, an estimate of the time taken for DNA to replicate itself when a cell divides is of the order of 10 hours. The take-away message is that chemical reactions are happening in unimaginably great numbers all the time and often incredibly rapidly.

The answer to Mary's question is: yes, they are indeed happening all the time: as you move, sit and even sleep. It is indeed extraordinary to contemplate the hive of activity proceeding at the molecular and cellular levels in our bodies – and in all other living things – of which we are normally quite unaware. Yet it is the sum total of these countless miniscule actions that gives rise to all our bodily experiences at the macro level: breathing, digesting, healing, growing, thinking and feeling. There is never complete rest from biochemical activity in the cells of the body; even asleep, when our bodies and minds appear to be at rest, we're still pretty active beneath the surface.

Nine

Enzymes seem to be talked about today much more than in the past. I remember as a youngster thinking of the word 'enzyme' as rather esoteric, connected in some way with curdling milk or brewing beer. Today, enzyme products are widely advertised, claiming, for example, to improve digestion and absorption of nutrients. 'Are they connected to digestive disorders?', asked Susan in a discussion one evening about digestion. 'I can imagine digestion as a set of chemical processes', added Mary. 'Food goes in in one form but comes out quite differently. But it's harder to imagine something like a muscle moving as due to chemical reactions'.

Watching a tub of milk shift from its swirling liquid form to solid curds in no time makes it plain that some kind of chemical change must be taking place and the added enzymes must be playing a part in this (Figure 9.1). This is what enzymes do: they bring about chemical changes (or reactions, as they are known). To be precise, they speed up chemical reactions that might otherwise be imperceptibly slow. This is what catalysts do, which is why enzymes are considered to be biological catalysts.

WHAT ARE ENZYMES?

You can see from the everyday examples we've mentioned why they are called 'biological' – they break up substances in milk to make cheese and, when added to washing powder, destroy patches of blood and food on dirty clothing. Enzymes interact with biological substances – like the proteins in milk or blood and the carbohydrates in grain. But, of even greater significance to us and our bodies, it is the role enzymes play, as the tools that carry out the everyday actions, that makes us tick. They break down the molecules of our food to give us energy, help decode our DNA, enable our tissues to grow and heal and play a vast number of other roles too.

Enzymes can be thought of as tools and, just as in a well-equipped workshop, there are many different types – thousands in the human

DOI: 10.1201/9781003272779-10

Figure 9.1 Cheesemaking.
(Image credit: US Department of Agriculture, Public Domain.)

body – each suited to a particular function. But in what way do they resemble tools – do they cut, drill and screw like tools in a workshop? 'Are they alive; do they die; do they reproduce?' asked Julie in the discussion group, perhaps thinking of the yeast or mother dough used in breadmaking. How do they carry out their many functions anyway? What are they actually made of – what is the equivalent of the metal, wood and plastic from which conventional tools are made?

Enzymes are not themselves living things. The yeast, bacteria and other organisms used in baking, brewing and fermenting are living things, but it's the inanimate enzymes within them that carry out the functions that are so useful to bakers and brewers. Enzymes themselves are very large molecules whose main role is to cut up or join together other molecules. Almost all biological materials from blood to nerves and muscle are made from molecules, and there are countless varieties of them. Enzymes perform their tasks by getting to the bonds that connect the atoms of a molecule and breaking them or, conversely, bringing atoms together and joining them up to make new molecules.

A molecule is simply a group of atoms held together by mutual attraction. Bonds are the links between the atoms that make up a molecule. Under normal condition, these hold firm – that's why our bodies, our furniture, our food stay roughly the same from day to day! But as we know, any of these can change over time – skin can age, furniture tarnish and food putrefy. These are manifestations of slow chemical reactions in which bonds between atoms in molecules get broken or made. Enzymes make chemical transformations happen much more quickly than they otherwise would – pretty important if you need to run from a bull, digest a meal or create a baby!

HOW ENZYMES WORK

The ability of enzyme molecules to cut and paste other molecules is down to the way they are structured. The bulk of an enzyme is a pretty firm structure that remains steady throughout any action it performs. A small part of it, however, is more flexible and, in the right circumstances, can move slightly – like the jaws of a pair of pincers or arms of pair of tweezers. This small part, called for obvious reasons the 'active site', is not only adjustable but is shaped very precisely to fit around just one specific type of target molecule.

It's this specificity that is so crucial to the work of enzymes. It means, in effect, that the active site of a given enzyme 'recognises' target molecules but is indifferent to all other types. This is represented diagrammatically in Figure 9.2. The red shape represents an enzyme molecule designed to split a specific smaller molecule into two parts. One example of such a process is the breaking up of long-chain molecules of starch in food into smaller units by the enzyme amylase which is present in saliva. Other kinds of enzyme in the stomach act similarly on the long-chain molecules of proteins in food.

The blue-coloured shape represents the molecule to be acted upon (often called the substrate). In our examples, this might be a stretch of a starch or protein molecule. The important point is that the active site of the enzyme molecule is specifically shaped to interact only with one particular type of substrate.

As the small substrate molecule docks into the active site, it triggers a small and momentary movement in the structure of the enzyme's active

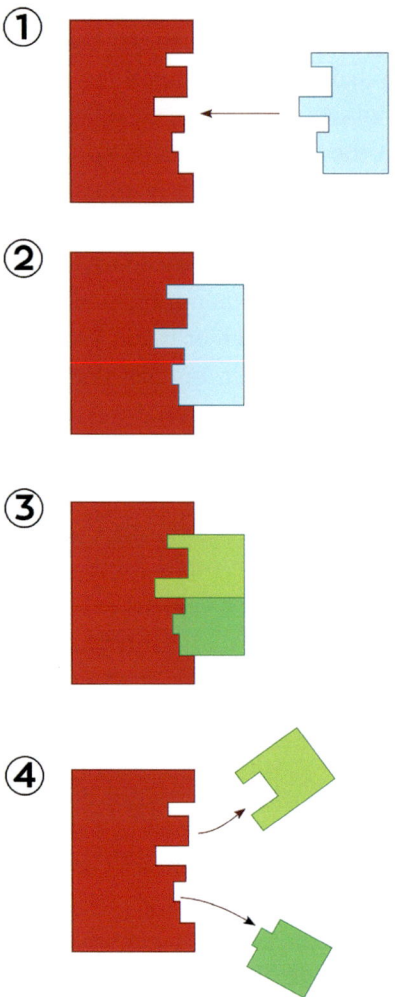

Figure 9.2 Diagram of an enzyme breaking up a molecule. (Image credit: domdomegg, Wikimedia https://creativecommons.org/licenses/by/4.0/deed.en)

site. As a result of this, one of the bonds between two adjacent atoms in the substrate is broken. The two pieces of the substrate separate and are released from the grip of the enzyme. The active site returns to its original state and the enzyme has completed its task of cleaving the starch, protein or whatever molecule. Sometimes other small molecules, such as hormones, may additionally affect the action of an enzyme, by subtly altering the main structure, away from the active site.

These kinds of enzyme – ones that break bonds between atoms – are crucially important for our bodies. As you can imagine, they play an especially important role in digestion. Our diets contain a huge variety of different substances – many types of protein, carbohydrates and fats for example. Most of the molecules in food are large and complex, often in the form of very long chains of sub-units. These need to be broken down into the smaller units of which they are composed, as described in Chapter 10. These smaller molecules are able to pass through the lining of our intestines into the bloodstream where they are circulated to where they are needed. A great menagerie of specific enzymes perform this cleaving role for us – and our fellow organisms.

Other kinds of enzyme play an equally crucial role in the opposite activity: building up new molecules from smaller ones inside cells. The smaller molecules produced in the digestion process are transported to cells all over the body where they are used to build the molecules we need: the proteins, carbohydrates and fats that make up our bodies (and those of other living things). These types of enzyme forge new bonds between atoms. An important example of this building process is the production of proteins in which sequences of small molecules called amino acids are linking together. Enzymes splice together one end of an amino acid molecule with the opposite end of another one, rather like the coupling together of coaches on a train. These illustrations of enzyme action – in digestion and protein production – are but two examples; enzymes are present throughout the body facilitating chemical reactions in almost every system of our bodies (and those of other organisms).

THE STRUCTURE OF ENZYMES

Before leaving this fascinating topic, we should take a look at what an enzyme is, as well as what it does. Perhaps the most remarkable property is its structure – the combination of a relative rigid backbone and a highly variable active site region. Enzymes are a particular class of protein molecule. They have blob-like shapes (technically 'globular') which result when the long thin chain of a protein molecule gets wrapped up. Like all protein molecules, enzymes are structured rather like a long necklace made of hundreds or thousands of beads (the amino acid sub-units). But rather than existing as a long thin thing, like a necklace around a neck, they fold into a compact shape, rather as a necklace does in the palm of a hand. Minute electrical attractions between some atoms in the chain keep various stretches of it in firm contact with other parts.

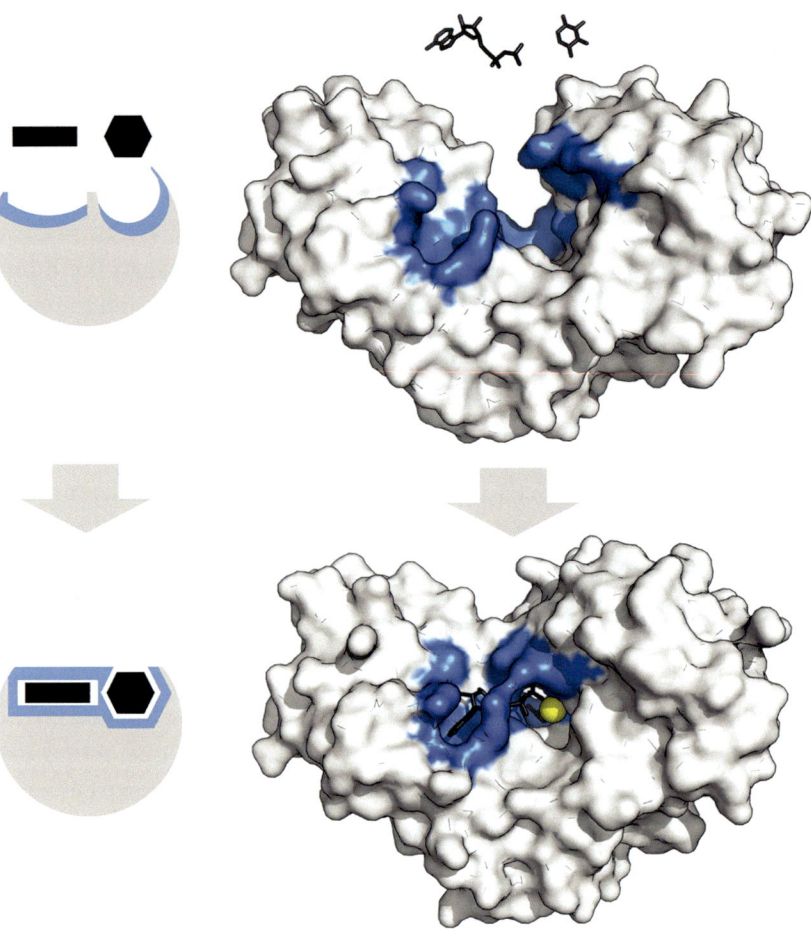

Figure 9.3 Model of an enzyme molecule (hexokinase).
(Image credit: Thomas Shafee via wikimedia, https://
creativecommons.org/licenses/by/4.0/deed.en)

The model of a typical enzyme molecule depicted in Figure 9.3 makes its overall globular shape clear. The knobbly surface represents the out-line of atoms. The active site, in blue, shows where two small molecules (known as the substrate) dock in and are joined together. Other kinds of enzyme split substrate molecules in two.

As an aside, many enzyme molecules also contain one or more metal atoms as part of the active site: for example, copper, zinc, selenium, cobalt and molybdenum. These metals are some of the essential micronutrients we need in our diet. Another aspect of enzyme structure is important in relation to food. In an overly acidic environment, enzymes can change shape and cease to function as a result. This is why pickling food items in vinegar can help to preserve them: it disrupts enzymes that might spoil the food. Similarly, chilling food can help preserve it by slowing down the activity of such enzymes.

CHEESE AND BEER

Equipped as we now are with an idea of what enzymes are, how they are constructed and what they do, let's look back at how they deploy their magic to make cheese from milk and beer from grain. Milk is a rich mixture of many types of chemical, including proteins, sugars, fats, minerals and plentiful water. To make cheese from milk, an enzyme capable of cutting up a particular protein is added. A common source of one such enzyme (rennet) is an animal's stomach, where it helps the animal digest milk. Vegetarian cheeses use plant-based enzymes instead.

Whatever the source, the enzyme has the capacity to break a bond between two atoms in one of the main milk proteins (caseinogen), resulting in a crucially different protein with similar name (casein). Molecules of the former dissolve happily in the watery mixture of milk, but those of its descendent, casein, do not. As a result, they precipitate out as a gel-like substance we call curd (see Figure 9.4). In a gel, the molecules

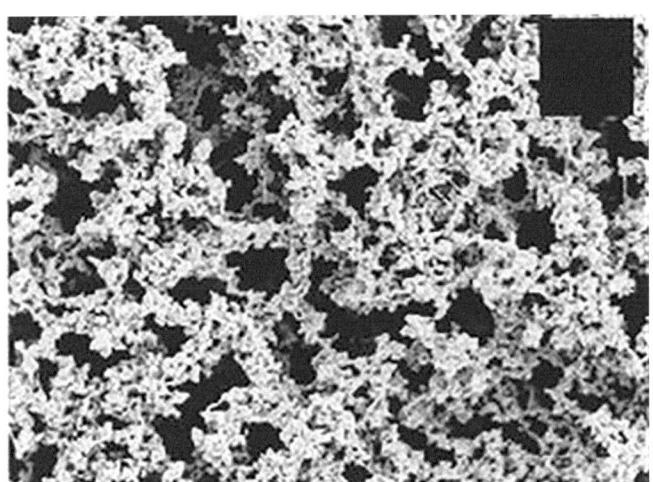

Figure 9.4 Gel produced by casein molecules (Scanning Electron Microscope image).
(Image credit: Peter Hristov, https://creativecommons.org/licenses/by/3.0/)

connect up in a kind of grid; this traps most of the fat molecules and calcium atoms in the milk. Together with other cheesemaking processes, the action of enzymes results in the separation of the curds from the watery whey, making the cheeses we know and love.

In the beermaking process, enzymes play a role in two different stages. In the early stage of mashing, naturally occurring enzymes in the grain convert some of the carbohydrates known as starches into smaller molecules, mainly sugars. Starch is a word used to describe a type of carbohydrate whose molecules are shaped as long chains of repeating units of the sugar glucose. In contrast to cheesemaking, the enzymes in beermaking break bonds between atoms of a *carbohydrate* rather than a protein – but the principle is much the same as for the breaking of bonds in milk proteins in cheesemaking.

In beermaking, however, enzymes from a different source play an additional role at a later stage – fermentation in which the alcohol is

Figure 9.5 Fermentation of beer.
(Image credit: Joshk via Wikimedia, https://creativecommons.org/licenses/by-sa/2.5/deed.en)

produced. These enzymes are not 'naturally occurring', but are normal components of yeast cells which are deliberately added to the sugary mix resulting from the earlier stage. Yeast is the name given to a particular type of tiny living cell. Like other living cells, yeast cells contain enzymes and some of these are able to break bonds in the molecules of sugars such as glucose and fructose. In the process known as fermentation, these enzymes catalyse the process in which sugar molecules are broken down producing ethanol (the common alcohol present in beer). Carbon dioxide is a by-product of this process; it forms the bubbles often associated with fermentation (Figure 9.5).

CONCLUSION

This exploration of the structure and function of enzymes has revealed something of what they are and what they do. It has focussed on enzymes involved in two particular processes – cheesemaking and brewing – in order to illustrate their role in familiar situations. Their ubiquity and importance throughout the living world, however, can hardly be exaggerated. They are critical to the everyday functioning of all living things, from yeast, bacteria and fungi to all forms of plant and animal life. They make the chemical reactions occur upon which all living processes depend. We humans have been exploiting their properties in food preparation since ancient times, long before they had even been identified, let alone understood. With our rapidly increasing knowledge about the precise roles they play, it's little wonder these once obscure actors are becoming the subject of everyday conversation.

Ten

'Can a low-carb diet really control type 2 diabetes?' asked Sarah in a science discussion after she had watched a TED talk to that effect. With the benefit of the internet, it's now relatively easy to find information about diet and type 2 diabetes as well as a host of other questions about diet, exercise and healthy living. The UK NHS website, for example, is clear about the most important message: a healthy diet balances a variety of types of food including 'fruit, vegetables and some starchy foods like pasta, but sugar, fat and salt should be kept to a minimum'. Diabetes UK cautions additionally that: 'the problem with some starchy foods is that [they] can raise blood glucose levels quickly, which can make it harder for you to manage your diabetes'.

Advice of this kind is helpful for most of us but it does introduce words that we may have only the slightest understanding of. What exactly is starch? How does it differ from carbohydrate? What counts as sugar? What are blood glucose and glycaemic index? In this chapter, we take the opportunity to explore some of the basic science behind these words from the dietician's lexicon.

GLUCOSE

Our bodies – and indeed all living systems – are the most marvellous examples of systems in balance. If there is too little oxygen, our cells begin to die; if there is too much, our lungs and eyes get damaged. Too high a temperature causes confusion, fatigue and ultimately death, but so too can too low a temperature. Bodily systems regulate this, restoring balance when levels stray either way. The thermostat in a heating system does much the same, causing a heater to click in or cut out depending on the temperature of the surroundings. The same applies to levels of a sugar known as glucose in the blood: too much leads to tiredness and nausea, too little to shakiness and irritation. In extremes, either can be life threatening. Regulation of the level of glucose is a vital task of the body. In diabetes, this regulatory system isn't working properly.

DOI: 10.1201/9781003272779-11

In science, the word 'sugar' refers not just to the single substance we might take with our tea, but to an entire class of chemical compounds with similar structures. Some, such as glucose, sucrose and lactose, are part of everyday talk, even featuring sometimes on food labels. Others, such as ribose and galactose, are less familiar. The common '-ose' ending has been adopted to define them as sugars.

Many sugars are sweet to taste. Evolution has developed our taste buds and brain circuits in a way that causes us to desire them strongly. This is believed to be a consequence of the advantage our distant ancestors gained from eating sugar-rich plant foods. This benefit arises from the high energy content of sugars; energy that is associated with the very structure of sugar molecules. The bonds that hold together the atoms in a sugar molecule are the source of energy, which can be released when the molecule is broken up. Energy gets stored in molecules as they are formed, much as energy gets stored in the spring of a jack-in-a-box as you close the lid. It was energy radiating from the Sun and falling on plants that originally led to the creation of such energy-rich molecules.

Releasing the energy stored in sugars involves an intricate process that occurs in most cells of our bodies – in muscle, nerve, bone and brain. Breaking down the great variety of large carbohydrate molecules in the food we eat into simple sugar molecules is one of the key roles of our digestive system. Following digestion, the energy-rich sugars are transported from the digestive organs around the body through the bloodstream. Molecules of one particular simple sugar, glucose (Figure 10.1), pass from the bloodstream into the cells in the body. Once inside the cell, glucose molecules are themselves broken down and the energy in them captured for later use. This occurs throughout the body – it provides the energy needed for us to move, digest and think.

DIGESTION

The food we eat may be very varied in appearance but the important molecules in it, from the nutritional point of view, are of four main kinds: starch, sugar, fibre and fat. Of these, fibre simply passes through the intestines without being digested, though it plays a key role in gut health generally. The other three kinds of molecule are digested, which means they get broken down into simpler molecules, mainly by the action of enzymes.

Starch comprises two kinds of molecule: amylopectin and amylose. These molecules are in the form of long chains. Each link of these is the

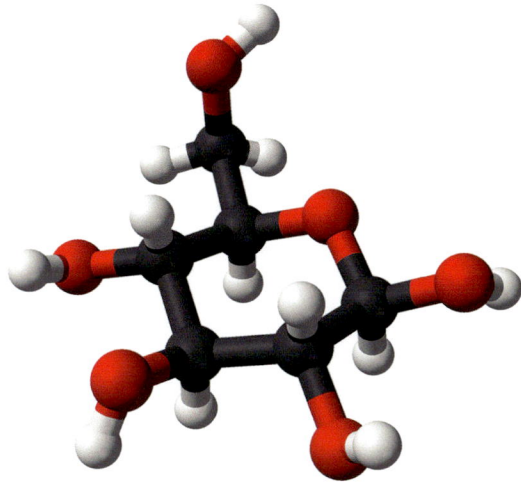

Figure 10.1 Model of a glucose molecule.
(Image credit: Ben Mills via Wikimedia, Public Domain.)

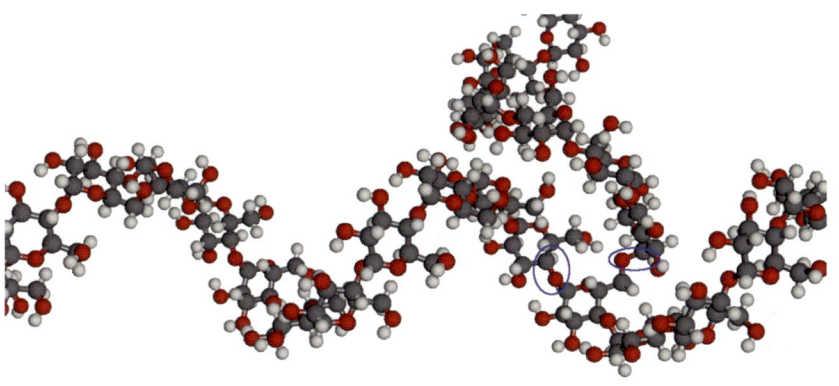

Figure 10.2 Model of the amylopectin molecule: a component of starch.
(Image credit: University of the West Indies Department of
Chemistry.)

much smaller molecule, glucose. The model in Figure 10.2 shows one
of these molecules: amylopectin. The molecules in starch get broken up
during the digestion process, producing the simpler molecules of glucose.

$$starch = glucose + glucose + glucose.....$$

The sugars contained in the plants we eat are of many types. The kind we
sprinkle on our breakfast cereal, sucrose, is not a long chain molecule

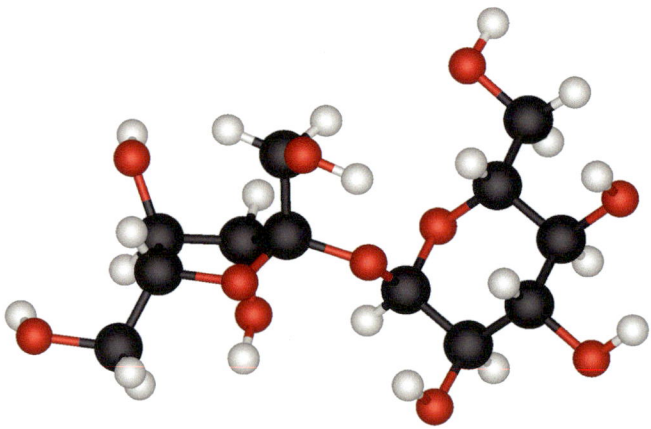

Figure 10.3 Model of a sucrose molecule.
(Image credit: Michael Ströck at en.wikipedia, https://creativecommons.org/licenses/by-sa/3.0/deed.en)

but is made up of two simpler sugars: fructose and glucose (on the left and right sides of Figure 10.3, respectively). Digestion breaks sucrose down into these two elementary sugar molecules.

$$sucrose = glucose + fructose.$$

Fat molecules (known as triglycerides) are also chain-type molecules (Figure 10.4). Their component parts are known as fatty acids. During the digestion process, these chains are also broken down, yielding the three separate fatty acids.

It is the glucose and fatty acid molecules that result from the digestion process that provide the energy needed for all our activities. After passing through the bloodstream and entering cells throughout the body, they are broken up releasing their stored energy.

REGULATION

As you digest your food, glucose and fats are released into the bloodstream. The fats are by their very nature oily molecules that are not soluble in a watery environment like blood. For this reason, they get bundled up inside a carrier called lipoprotein that is soluble. The two kinds of molecule – glucose and fats – convey the energy needed by your cells but have different qualities. Because of this, the two are able

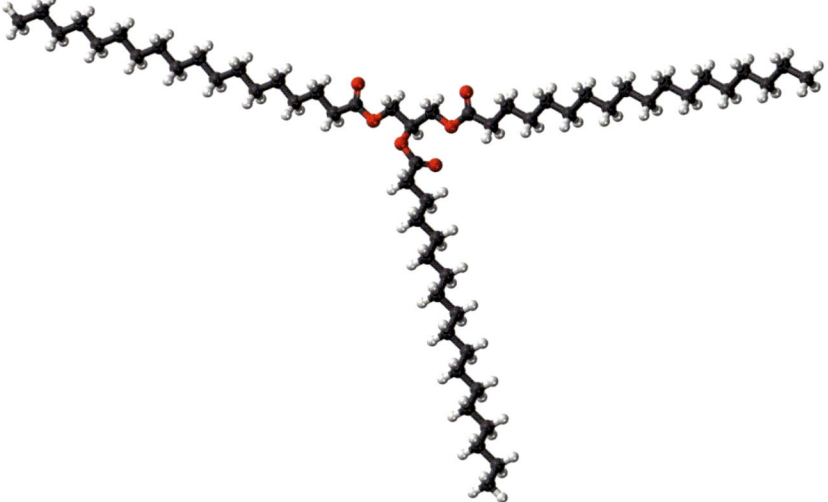

Figure 10.4 Model of a triglyceride ('fat') molecule made of three fatty acids.
(Image credit: Jynto via Wikimedia, Public Domain.)

to work in complementary ways to ensure your cells get the supply of energy they need all day long, including between meals. Your immediate need for energy is satisfied by the glucose molecules which are delivered straight from the bloodstream to all the cells of your body. For energy needed between meals, fatty acids are held back. They are stored in the so-called 'fat cells' that exist in various parts of the body and in the liver.

For glucose molecules to break into cells, however, they need help. The membranes that surround cells are water resistant – that's how cells keep their interiors separate from what's around them. Glucose molecules can't pass through them unaided. The help they need is provided by molecules of a hormone called insulin. Insulin molecules dock onto large protein molecules lying on the surface of cells, known appropriately as insulin receptors. This has the effect of opening up 'pores' in the cell wall, allowing the glucose molecules to pass through into the interior of the cell where they are able to release their energy.

To enable this to happen, insulin has to be released from the pancreas where it is stored while you are digesting your food. As you digest your food, glucose levels begin rise in the bloodstream. In a neat piece of signalling, this gets detected and triggers the release of insulin molecules from the pancreas. Insulin serves many functions; in addition to opening up all your cells to take in glucose, it signals to fat cells and the liver to take in and store up molecules of fat from your food for later use. In between meals, as the glucose level in the bloodstream begins to drop, your insulin level drops with it. As a result, fat cells cease storing up fat but instead release it into the bloodstream, keeping up the supply of energy to all cells. Fat and glucose work together to keep you fuelled. That's why you need both fats and carbs in your diet.

DIABETES 2

Problems arise, however, if your insulin system ceases to work properly. It can become less effective in helping glucose into your body's cells and cause the liver to over-produce glucose that isn't needed. This can lead to an excess of glucose floating around in your bloodstream. This is what doctors are looking out for when they order a blood glucose test and ask you to fast beforehand. It is a sign of type 2 diabetes.

Like all aspects of our bodily systems, the digestive processes we have today evolved very slowly over countless millions of years. The system we have today suited early humans, but doesn't work so well for us today. Back at that time, carbohydrates in food were encased in fibre, making them relatively slow and hard to digest. Sugars were more of a rarity then, too. Today with fibre stripped away and sucrose added to so much of our food, digestion is a much easier and faster process. As a result, as we eat, glucose can flood into our bloodstream too fast for the balancing act that insulin normally maintains. Excess insulin enters the bloodstream in an attempt to cope. But our cells can become overwhelmed; just not able to process all the glucose coming their way. Fat cells may also be in a similar situation: unable to cope with the excess fat coming their way. As a result, half-processed glucose and fat and waste products may abound when too much low-fibre, high-sugar food has been consumed.

To cope with a crisis of this kind, our body's defence mechanism – the immune system – gets called into play. In an attempt to clear up the mess, it releases chemicals that reduce the effectiveness of insulin. It's trying to protect our overwhelmed cells. As a result, less glucose makes it into our cells and less fat gets stored in the fat cells. It has to go somewhere, so it stays floating around in our bloodstream. That is where the danger lies: too much partially processed glucose and fat in the bloodstream. If such overloading becomes persistent, a serious threat is posed for the heart and liver: that's type 2 diabetes. The body handles imbalances of this kind satisfactorily when they occur from time to time – as we know from our occasional moments of excess, such as religious festivals. It is the regular and excessive consumption of sugary, low-fibre foods that pose the risk.

DIET

There's a lot more to healthy eating than we can cover here – the role of fructose (one of the breakdown products of eating sugar), the role of 'good' bacteria in our gut and the effect of additives, for example. On top of this, there is the part the hormones – leptin and ghrelin – play in telling our brains whether or not to feel hungry and the psychological tendencies that can drive us to compulsive or even addictive habits.

For practical purposes, however, research has thrown up some clear and consistent messages about how to give our insulin a chance and avoid type 2 diabetes. The Harvard School of Public Health summarises these and offers the encouraging conclusion that, in the vast majority of cases, prediabetes and type 2 diabetes can be avoided by making simple lifestyle changes. Its suggestions include the following:

- Exercise: working your muscles more often and making them work harder improves their ability to use insulin and absorb glucose
- Choose whole grains and whole grain products over refined grains and other highly processed carbohydrates
- Skip sugary drinks, and choose water, coffee or tea instead
- Choose healthy fats. polyunsaturated fats found in liquid vegetable oils, nuts and seeds can help ward off type 2 diabetes. Trans fats do just the opposite.
- Limit red meat and avoid processed meat; choose nuts, beans, whole grains, poultry, or fish instead.

These are manageable, evidence-based tips that can be borne in mind as you choose what to eat from day to day. If you are looking for a specific diet, however, the evidence suggests that simply following a dietary fad often doesn't work over the long term. Losing a few kilos early on may prove relatively easy but over the long term most people find it very hard to sustain a diet. The desire to eat sweet and fatty thing is a very strong bodily urge, more or less hard-wired into the human brain. Evidence-based commentators generally advise us to take dietary changes slowly, allowing ourselves some pleasures as we go along, reminding ourselves from time to time how hard it is to overcome our built-in impulses. It may prove a bit of a struggle but the health benefits make it worth the effort.

Eleven

'Don't get me started on glue' – a rather unexpected request during a conversation with a friend. Apart from her fascination with Pritt sticks, she'd once had the misfortune to get stuck by the roadside after having tried to repair a damaged wing mirror with superglue. Her enthusiasm for adhesives, however, had developed out of admiration for the wonders of glue technology: adhesive strips strong enough to keep a shelf in place and Post-it® notes that stick, yet peel apart.

The topic springs to mind in so many everyday situations yet rarely makes it into the realm of popular science. Adhesion is so ubiquitous that it's like the air we breathe; you scarcely notice it although it plays a huge role in our everyday lives – from cleaning our teeth to decorating the bathroom. Is it really as dull as it at first seems? Advances in technology have led to remarkable innovations in recent decades: superglue in the late 50s and the indispensable Post-it® in the 1970s, for example. But glues also have a long history, serving us for millennia by holding our furniture together. Cement has been doing much the same for our buildings – and, more recently, our teeth.

An even more extraordinary aspect of adhesion is the capacity to hold things together in the absence of either glue or cement: daddy-longlegs on bathroom tiles, sheets of glass that refuse to come apart. What is it that makes one substance stick to another? Even more fundamental: what is it that keeps substances separate from one another in the first place; after all, sand and cement get along perfectly well together in their separate ways until you add water? Come to think of it, why does stuff hold together at all? A puff of smoke soon disperses but a lump of coal doesn't. Let's start by clarifying a few basics.

MATERIALS

Most of the materials we come into contact with are solid – metal, plastic, cement and wood, for example. If you were to look at them in detail through a microscope, you'd see they are each made up of microscopic

DOI: 10.1201/9781003272779-12

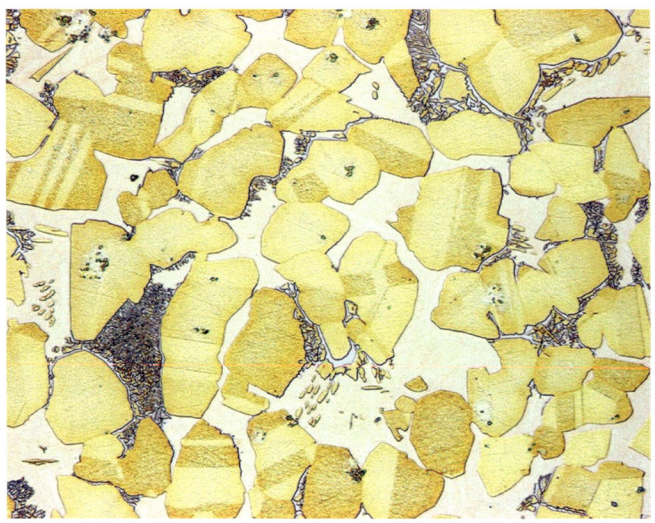

Figure 11.1 Grains in a metal.
(Image credit: Eisenbeisser, https://creativecommons.org/
licenses/by-sa/3.0/deed.en)

structures at a more granular level. Metals, for example, consist of grains, as the microscope photograph of grains in an aluminium-bronze alloy in Figure 11.1 shows.

Plastics, however, are made of polymer molecules – i.e. long chains of repeating links. These come together to form regions that are semi-crystalline, known as spherulites.

The photograph of polypropylene in Figure 11.2, taken through a microscope, shows these semi-crystalline structures.

Cement used in building is also composed of grains when seen under a microscope. The microscope photograph in Figure 11.3 is of a limestone-based cement.

A different kind of cement is used to hold bridges and crowns in place in dentistry. Though made of different substances, it also has a granular structure, as shown in the microscope image in Figure 11.4.

Wood, on the other hand, as a biological substance, is made of plant cells, each with a complex internal structure. Within the walls of these cells are fibres made of several distinct components. Long thin molecules of a protein called lignin line up with microfibres of carbohydrates called cellulose and hemicellulose with which they interconnect (Figure 11.5). It is these that give stiffness to the cell walls and hence the wood itself.

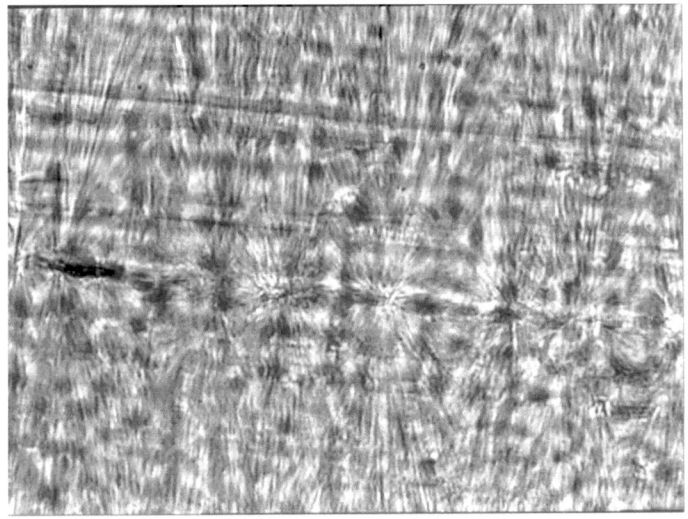

Figure 11.2 Semi-crystalline structure of polypropylene.
(Image credit: Unbound via Wikimedia, Public Domain.)

```
Acc.V  Spot Magn    Det  WD  Exp                    50 µm
15.0 kV 5.5  500x    BSE 10.1  1      0.3 Torr M010704
```

Figure 11.3 Limestone cement seen through a microscope.
(Image credit: Zhijun Tan (土木坛子).)

Figure 11.4 Dental cement under the microscope.
(Image credit: *Journal of Biomedical Science and Engineering*, Open Access, http://creativecommons.org/licenses/by/4.0/)

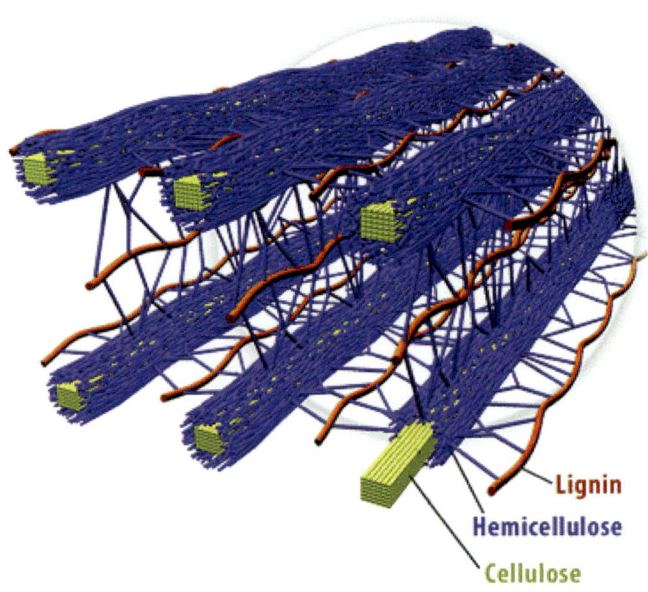

Lignin
Hemicellulose
Cellulose

Figure 11.5 Diagram of the microstructure of wood.
(Image credit: U.S. Department of Energy Genomic Science program, https://genomicscience.energy.gov)

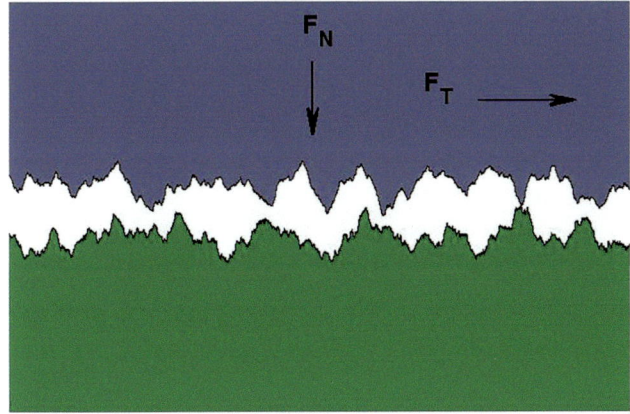

Figure 11.6 Diagram of two surfaces in contact.
(Image credit: CaoHao of matlab, https://creativecommons.
org/licenses/by-sa/4.0/deed.en)

What each of these materials have in common, whether granular, crystalline, cellular or fibrous, is that their interiors have an irregular structure which leaves their surfaces rough at the microscopic level. As a result, when two surfaces come into contact with one another, they are less intimately connected to one another than appears to the naked eye. Figure 11.6 gives an impression of how it might look at the microscopic level. The jagged surface of the upper object sits irregularly on the jagged surface of the lower one. It's only at very few points that the surfaces physically touch each other

Incidentally, this image also shows why it takes some force (F_T) to drag one material across the jagged surface of another one – what we know as friction. Imagine those peaks and troughs grating as they pass each other. The arrow F_N also indicates that, if we press one object down on top of another, the jagged surfaces mesh more tightly together, making it harder to slide one over the other – increased friction. The role of adhesives is made immediately apparent by this diagram: their role is to fill the space between two surfaces, connecting them together much more intimately than they would otherwise be.

ADHESIVES

The microscope photograph in Figure 11.7 shows beautifully how an adhesive, in liquid form, penetrates and fills the gap between two surfaces. It was taken by a scientist at the government forestry research

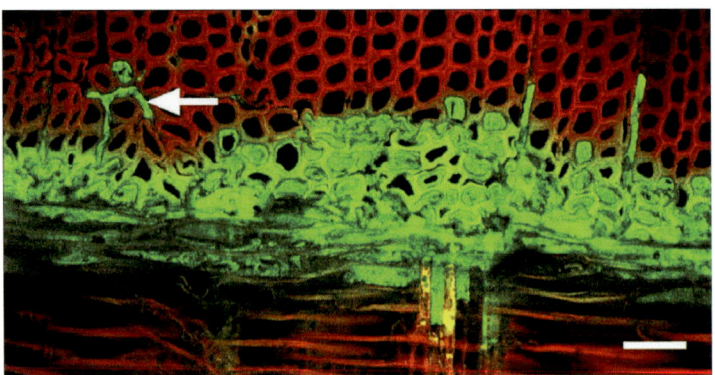

Figure 11.7 Adhesive between two plies of wood.
(Image credit: Bernard Dawson of Scion, NZ.)

institute in New Zealand. It shows a line of adhesive (green) between two sheets (or plies) of wood (red). The adhesive can be seen penetrating deeply into the cellular structure of the wood on either side. To indicate the scale, the white line at the bottom right represents 50 micrometres (millionths of a metre).

Adhesives work their way between surfaces to perform their binding role in a variety of ways. The diagram in Figure 11.8 illustrates four of them. In the simplest case (top left), molecules in the adhesive – yellow in the diagram – and the two surfaces may simply attract one another – *adsorption*. Alternatively, molecules in the adhesive may react chemically with those on the two surfaces – *Chemisorption* (top right). In other cases, molecules may cross between the materials (*diffusion*) or penetrate the cracks in them, linking them physically (*mechanical adhesion*).

It's clear from each of these images that the adhesive needs to spread evenly throughout the gap between the two surfaces – it needs to 'wet' them. This fits with our everyday experience of glues as liquids, gels or pastes. But to become firm and fast after having wet the surfaces, they also need to dry out over time. To achieve this, the adhesive is often combined with a solvent that evaporates, such as water, acetone or ether – that's the smell of glue setting. As well as sticking firmly to other materials, glues also need to hold together strongly within themselves, so that the joint doesn't come apart within the glue itself. In other words, it must cohere as well as adhere.

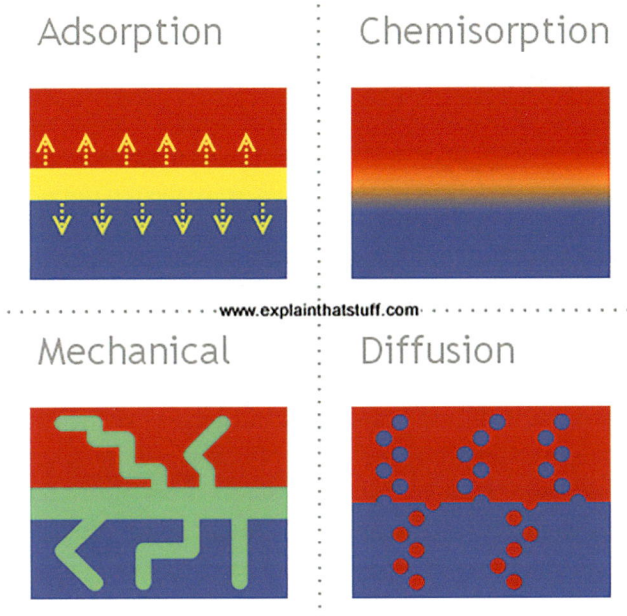

Adsorption · Chemisorption

www.explainthatstuff.com

Mechanical · Diffusion

Figure 11.8 Four ways in which adhesives bind two surfaces.
(Image credit: Chris Woodford, explainthatstuff.com)

THE ATTRACTIVE FORCE

Filling the space between two surfaces and perhaps penetrating into them is clearly necessary for glues to work, but what is it that actually does the sticking?

The explanation lies not in some special 'stickiness' substance but in a simple but fundamental aspect of all matter: the electrical nature of atoms. It had long been thought that all the diverse substances we see are made up of tiny, fundamental components, to which the ancient Greeks gave the name 'atoms'. What was not suspected till atoms were actually discovered, just over a century ago, was that these are electrical in nature. This seems to be counterintuitive as most materials appear to be perfectly neutral in most normal circumstances – neither positively nor negatively charged. The surprise was to find that atoms are not in fact fundamental, indivisible units as originally expected: they are themselves made of smaller parts and it is these that are electrically charged. The trick is that the number of positive and negative particles inside atoms is exactly balanced; that's how they come to be neutral overall.

For adhesive substances, it's the electrical attraction between positive and negative charges that accounts for the bonding. Molecules are made of atoms and, like most substances, glues are made of molecules. Inside some molecules, despite being electrically neutral overall, one part can be a bit more negative than another, leaving other parts a bit more positive. Where a molecule like this meets another such molecule, the positive and negative regions (or poles) can be attracted to one another much as the opposite poles of two magnets are.

These forces of attraction between two molecules are pretty weak in themselves and get weaker the further apart the molecules are. They add up, however, when many molecules come close to one another. In the interior of solid and liquid substances, molecules are close to one another. It's this electrical attraction that holds molecules together, making a substance cohere. It's what makes a piece of wood or chunk of stone a thing, rather than a vaporous wisp. When two objects come into contact, however, relatively few of the molecules on their surfaces come close to one another, due to the roughness of typical surfaces. That's why most substances don't ordinarily stick to one another: the overall electrical attraction between molecules at their surfaces is too weak.

As we have seen, adhesives fill up the gaps and ensure that the molecules of which they are made get right up close to those in the material they are sticking to. The strong bond that develops is the result of a huge number of tiny forces adding together.

For glues to work, their molecules have to be attracted to the molecules of the two substances being joined together. In Figure 11.9, the blue molecules need to be attracted to the red ones. This property is known, unsurprisingly, as adhesion. In addition, however, the glue itself must not split apart internally. The glue molecules themselves (in blue) must also be strongly attracted to one another; there must also be cohesion.

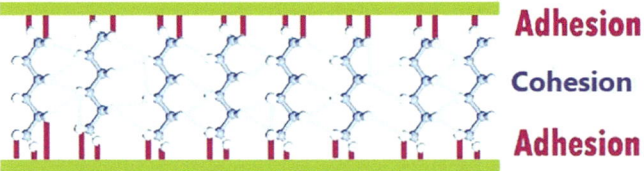

Adhesion
Cohesion
Adhesion

Figure 11.9 Adhesion and cohesion at the molecular level.
(Image credit: Aldobenedetto Zotti.)

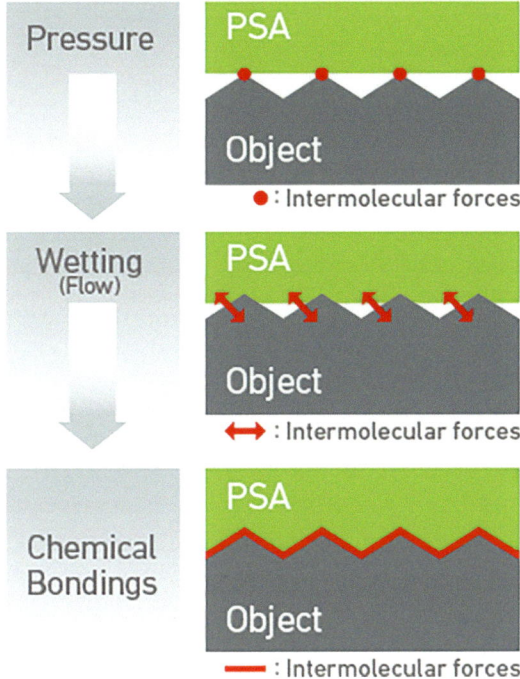

Figure 11.10 Mechanism of action of sticky tape.
(Image credit: DIC Corporation, Tokyo.)

EXAMPLES

Now we understand the principles, what about the practice? Glues and cements appear in all sorts of situations, from bookbinding to dentistry, wall paint to insect feet. Some examples from everyday life are explored in this final section.

Sticky Tape

Formally known as pressure-sensitive adhesive (PSA), sticky tape combines a viscous adhesive (often a polymer) with a plastic film on which it is fixed. The backing of the film has to be coated with another substance to prevent the tape from sticking to itself. Gently pressing sticky tape with your hand helps the adhesive to flow, as illustrated in Figure 11.10. This allows the tape to make closer contact with the surface, increasing the amount of force between the molecules. If dust or oil gets in between, it reduces the contact area, weakening the bond.

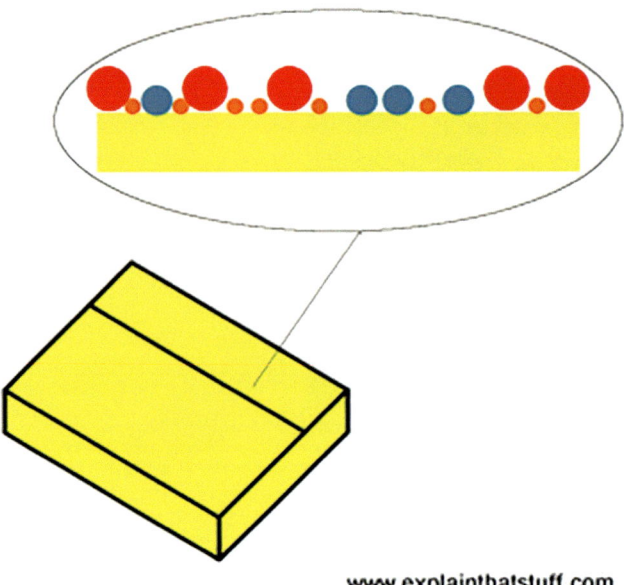

www.explainthatstuff.com

Figure 11.11 Mechanism of action of Post-it notes.
(Image credit: Chris Woodford, explainthatstuff.com)

Post-it® Notes

Post-it® notes are known as 'low-tack' pressure-sensitive adhesives. This means they do not wet or run rapidly like other glues. The sticky part contains tiny bubbles of adhesive, so when pressed onto a surface there's just enough adhesive force to grip lightly (Figure 11.11). At first, the 3M corporation considered the accidental discovery of a new weak glue to be a failure. Years later, an employee started using it to fix bookmarks in a book; some market testing soon led on to commercial success. The story goes that only yellow paper was available in the lab at the time to test it on – that colour has stuck.

Glue Stick

This innovative type of glue was developed in 1967 by researchers at Henkel in Düsseldorf. The idea of a glue stick was a departure from the usual liquid type. To get the glue to stiffen, a soap gel was combined with the water-soluble adhesive, making it solid. When rubbed on paper, the adhesive components are released creating an adhesive film. Lipstick was the inspiration for the twist-action. Apparently, the name for the original brand came from the inventor's child mispronouncing 'pretty stick'.

Superglue

Formally known as cyanoacrylate, superglue works differently from most other glues. Instead of using a solvent that evaporates to leave a firm bond, the molecules of cyanoacrylate rapidly fuse together in the presence of water, forming long chains (or polymers). Even the water in the atmosphere or on your skin is enough to set this off. The polymers then bond strongly to the molecules in the surfaces to be glued. The discovery of superglue was another example of serendipity. It began during the Second World War in an effort to find a clear plastic for gun sights. The research was abandoned when the substance turned out to be annoyingly sticky. It was accidentally rediscovered later by researchers at Eastman Kodak who foresaw its commercial potential.

Dental Cement

As with superglue, dental (or 'luting') cement does not involve a solvent drying off. Instead, a chemical reaction takes place between a dry and a wet ingredient. As the cement hardens, it bonds with the two surfaces. Sometimes, halogen or LED light is used as a catalyst to speed up the reaction.

Cells

In living things, a fundamental kind of adhesion is the stickiness that holds together the cells of which they are composed. Without this, you or I would just be a heap of powder rather than a complex of connected tissues. The 'glue' that connects cells comprises adhesive molecules that create junctions between cells (Figure 11.12). They are not only tight – so that tissues hold together even under strain – but also extremely flexible, as we know every time we stretch or even breathe. An adhesive molecule (called cadherin) links through a protein to the internal structure of a cell. It's recently been discovered that these proteins are capable of assembling, disassembling and reassembling thousands of times in a second. It's this rapidity with which connections between cells can be made and broken that enables our tissues to be flexible yet hold together firmly.

Geckos

Geckos are able to run around upside down thanks to thousands of tiny hair-like bristles on the underside of their feet (Figure 11.13). Even smaller projections are attached to these. As we've seen with all other adhesives, it's this multiplication thousands of times that creates

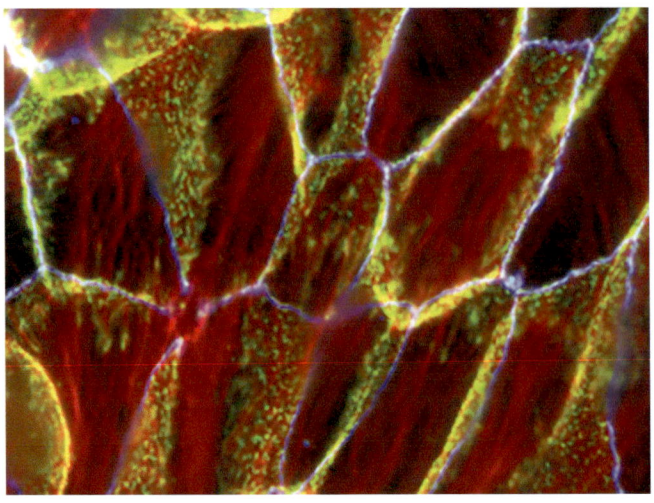

Figure 11.12 Adhesion between cells (blue lines).
(Image credit: Sergey Troyanovsky, Northwestern University.)

Figure 11.13 Underside of a gecko's foot.
(Image credit: Bjørn Christian Tørrissen, https://
creativecommons.org/licenses/by-sa/3.0/deed.en)

sufficient attraction to hold surfaces together. Geckos, like Post-it® notes, need to also detach their grip rapidly as they move. The bonds between the hairs on their feet and the surfaces on which they scamper are easily broken.

The gecko's anatomy inspired scientists at Stanford University to develop a device enabling humans to climb a vertical glass surface (Figure 11.14). By assembling thousands of microscopic wedges of silicon rubber onto tiny tiles set onto pads, they were able to generate adhesive forces just as the gecko does.

Figure 11.14 Inspired by geckos.
(Image credit: Eric Eason, Biomimetics and Dexterous Manipulation Lab, Stanford University.)

CONCLUSION

What this brief exploration of adhesives reveals is that a simple and unifying principle underlies the great diversity of apparent types. It's hard to imagine that the glue holding the ultra-hot tiles on a returning space capsule works on the same fundamental principle as that on the back of a Post-it® note. Of course, the formulation differs but the underlying mechanism remains the same: attractive forces between molecules. Being so weak individually, we hardly notice them in ordinary circumstances; plates don't stick to tables, handshakes are easily released. The job of any adhesive is to increase dramatically the number of molecules in close contact between two surfaces so these tiny forces multiply up to a significant grip. A simple look at an everyday phenomenon has once again led us to a foundational concept in science: the electrical nature of matter.

Twelve

Call me a geek, but I can't help thinking of my school science lessons every time I take a shower. It's the problem of the mixer taps. What is it that actually comes out of the cold and the hot one ... apart from water?

'Presumably heat comes out of the hot one, but what about the cold one?' asked Harry in a discussion one day 'does cold come out of it?' 'What actually do we mean by heat anyway?' Jean queried. 'What about temperature – what do hot and cold really mean ... they are pretty subjective, aren't they?' added Helen.

Back then, as a 14-year old, it all seemed so confusing to me. In a science practical, we once had to make a thermometer out of glass tubing and suck up some red liquid. I distinctly remember the ever-patient Mr. Sills trying to explain the strange word 'calibration' as we plunged our glass-tubes into bowls of ice, scratching marks on the glass as we did so (Figure 12.1).

Many decades later, I can now see there's a fundamental difficulty caused by confusion in the very language we use. 'Hot' is an adjective, 'heat' a noun. It's a pity they sound so similar, given they play such different roles in language and have such distinct meanings. Then there's the opposite; 'cold' is the adjective and ... what is the noun: coldth? coldness? Unfortunately, we tend to use the same word for both: 'It's cold in here. Shut the window, you'll let the cold in' – ambiguity in language hampering our efforts to understand.

The confusion partly dates back to a discredited theory, popular in the eighteenth century. Back then, long before the advent of steam locomotives and central heating, scientific efforts to understand heat gave rise to the idea that it was a kind of fluid. After all, it appears to flow along a teaspoon in a teacup from the hot end to the cold, just like rainwater flowing from the higher to the lower end of a gutter. Steeped as science was in the classics, this supposed liquid was named 'caloric', after the Latin word for heat.

DOI: 10.1201/9781003272779-13

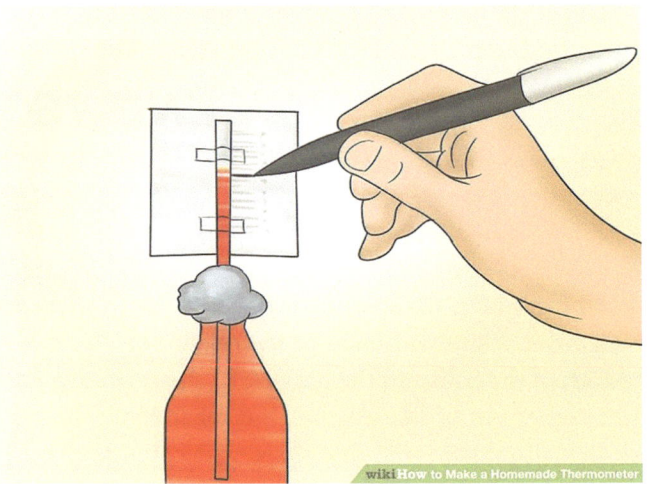

Figure 12.1 Calibrating a home-made thermometer.
(Image Credit: Wikihow, Public Domain.)

The development of the new science of thermodynamics in the nineteenth century was driven by the arrival of steam engines, used originally to pump water out of mines which tended to flood as they were dug ever deeper. The new technology inspired new science which resulted in an altogether different and powerful concept of heat. This is broadly how we understand it to this day. It isn't any kind of extraneous thing added into stuff, not a fluid running through the metal of a teaspoon. It's simply an intrinsic property of the stuff of the teaspoon itself.

Fortunately, scientific understanding of the nature of 'stuff' was also developing around the same time. This led to the concept of what we now know as atoms and molecules (groups of atoms bonded together). All materials – solid, liquid and gas – are composed of these. What was soon realised is that these atoms and molecules aren't just sitting quietly together, they are incessantly moving.

They may be dashing about freely and separately like midges on a summer's day – that's how it is in gases; or they may be more closely packed, sliding past each other, like balls in a children's ball-pit – that's the liquid state. In a solid, they are lined up next to one another like eggs in an eggbox but they are still in active motion, vibrating about their fixed position. Heat, it was soon realised, is simply the net effect of all this movement, summed up over the zillions of separate atoms or molecules that make up any piece of matter.

HEAT AND TEMPERATURE

This more enlightened concept, in which all the atoms or molecules in a thing are constantly jiggling around or shifting, clarified what we mean by heat. If they are moving, they must have energy – the faster they move, the greater the energy. 'Heat' is simply a reflection of the energy atoms and molecules have due to their motion. The 'heat' in a bath or an animal or a star simply means the sum total of all the energy of all the atoms and molecules moving in it. As a bath fills up, more and more water molecules are being added, each with its own energy of motion; so the total amount of heat energy is increasing. It's this energy of all the H_2O molecules that you're paying for through your energy bills.

But there's another catch – your bath could be warm or hot. You pay more for a bath full of hot water than the same amount of warm water. Temperature is a measure of how hot something is and a number of scales have been created to quantify this. The widespread Celsius scale, for example, uses the properties of water for this, assigning the value $0°$ to the temperature at which it freezes and $100°$ to its boiling point. The calibration experiment I recall from school involved us making marks on a glass tube at these two temperatures and drawing a scale in between.

You might well think it's somewhat circular to say temperature measures how hot something is. It doesn't really give us clear picture of what temperature actually is. Fortunately, as it became clear, during the nineteenth century, that molecules were in a constant state motion, the concept was put on a firmer footing. Temperature is a reflection of the *average* energy atoms and molecules have due to their motion. In any given thing – solid, liquid or gas – molecules are moving to differing degrees: vibrating, rotating or shifting around. It's the average energy of this movement that determines the temperature. The greater the average energy, the higher the temperature.

A spark from a sparkler is obviously 'hotter' than a warm bath, its temperature is higher – it could burn you. In a warm bath, by contrast, there's much more stuff – many more molecules – in comparison to a sparkler – but, on average, each molecule has relatively less energy compared to those in a spark. So there we have it – the basic concept of thermodynamics. Heat is a form of energy and it represents the total energy of motion of all the molecules making up a thing. Temperature, on the other hand, is a measure of how much energy is associated with each molecule, *on average*: less energy at cooler temperatures, greater energy at higher ones. A bath contains lots of less energetic molecules; a sparkler a smaller number of more energetic ones.

So when you take a shower, you balance water from the hot and cold taps to give you the right temperature. If the temperature is too low, you can increase the flow of high energy molecules by opening the hot tap. This raises the proportion of high energy molecules in the mix, thus raising the temperature. Equally you could reduce the proportion of lower energy ones by closing the cold tap. Either way, you're raising the *average* energy of the molecules in the mix and hence raising the temperature.

Heat on the Move

Switching our attention from a shower to a bath throws up another obvious, yet fundamental aspect of heat energy. It won't stay where it is for long; heat moves, baths cool down. If you fill a bath from separate hot and cold taps, you'll notice it takes time for the heat energy from the two taps to spread out evenly throughout the water. You may physically stir up the water to hasten this process, but even if you don't, a bath filled with hot and cold water will eventually settle, unaided, to a steady temperature in between the two. Over time, it will cool down, the temperature gradually dropping as it loses heat energy, even though no water is physically draining away.

With our understanding of heat as simply energy due to the motion of molecules, we can now understand more clearly what is happening as the temperature evens out in our bath. Energy is gradually transferring from higher temperature to lower temperature zones. There are several different ways in which heat energy can move: you may dimly recall a mantra about this from science at school: 'convection, conduction and radiation'.

In a liquid, like the water in a bath, or a gas, like the air in a room, the warmer zones will tend to rise upwards as they are a little less dense than the cooler parts. As they do so, cooler water or air will move in lower down, to replace what had moved upwards. This process sets up a circulating current of the liquid or gas: a process known as convection. These currents help to distribute the heat energy from the warmer to the cooler regions (Figure 12.2).

Other processes are also at work. At the surface of a bath full of water, some heat energy will simply radiate away, much as the energy radiates

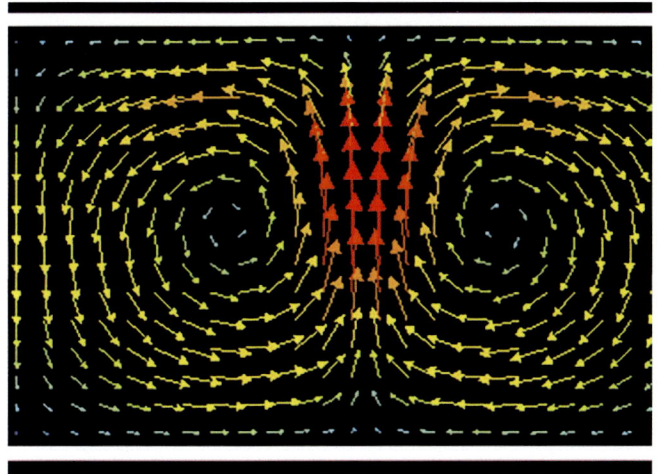

Figure 12.2 Convection currents.
(Image credit: Harke via Wikimedia, Public Domain.)

from the sun or an electric fire. This means of transferring energy, known as radiation, is similar in kind to the way energy associated with light or X-rays moves. The energy simply travels outwards in a straight line through space, even if there is no air or water to carry it; it travels through a vacuum. Known as infra-red, this type of radiation is what is picked up by infra-red cameras used to detect animals at night or heat loss from buildings.

The third, and perhaps most straightforward means by which heat can transfer is by conduction. As its name implies, the energy is simply conducted outwards from a high temperature region to a lower one by direct transfer of energy from one molecule to its neighbours. For example, when heat flows along a teaspoon from a hot cup of tea to your hand, energy passes from the more energetic molecules in the tea to the less energetic ones in your hand. It's not anything concrete that moves along the teaspoon – not the molecules themselves, nor some imaginary 'caloric' fluid flowing between them, but simply the energy of movement passing from more agitated molecules to less agitated neighbouring ones (Figure 12.3).

An animated version of the conduction process can be seen at http://taylorsciencegeeks.weebly.com/uploads/5/9/2/0/59201005/364918302.gif

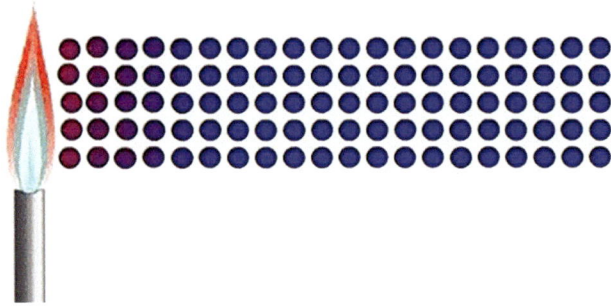

Figure 12.3 Agitation of molecules in heat conduction.

So, to nail these concepts of heat and temperature, let's look more closely at some homely examples of them. Your shower is pouring out hot water at a steady rate. Zillions of water molecules are rushing past, each carrying their bit of energy. You are paying for this through the gas or electricity consumed in adding energy to your hot water in the first place – via your boiler or heater. When you adjust the temperature of your shower, you are deciding how much energy these molecules will have on average – the temperature. The higher the temperature, the greater the energy of the molecules; so, of course, the more energy you consume and the more you pay in bills. You'll also pay more, if you increase the flow rate of the shower. This time it's not because the molecules are more energetic, but because more of them are passing each second.

Now let's take a hot cup of tea. The molecules in it will be moving with a range of speeds, but on average they will have greater energy than those in a cool teaspoon placed in the cup. As you can imagine, the greater energy of the molecules in the hot tea will gradually transfer to the molecules in the spoon, passing from one agitated molecule to the next. This elevated level of energy will spread gradually along the spoon to your hand holding it. Of course, it only spreads from the more energetic tea to the less energetic spoon – not the other way round! It's easy to see how this corresponds to what we know intuitively about temperature – unaided, heat will only pass from a higher temperature to a cooler one, not the other way around.

CONCLUSION

These simple ideas about heat and temperature help with understanding many aspects of everyday life. When we insulate our homes to reduce heat loss, we are impeding the flow of energy, through brick, wood or glass, from the warmer interior to the cooler outdoors. When we close the door of a fridge, we are creating an enclosed box from which heat energy in the food, air and walls can be extracted. Having a clearer picture of what is actually meant when we talk loosely about heat is not only enlightening, it can also help us reduce our bills and save our planet.

What Popping Tells Us about Pressure, the Middle Ear and the Atmosphere

Thirteen

One evening, deep into a cold November, a fascinating science discussion kicked off, inspired by Julie's trip to the Canary Islands. Her flight to the sunny south had started badly with a typical British cold: blocked ears and nose. As the plane ascended however, she was delighted to find her ears becoming unblocked. Her sense of relief lasted a pleasant hour or two before she sadly found them blocking up again as the plane descended into Fuerteventura.

She guessed it must have been to do with the pressure in the cabin. This simple suggestion triggered an avalanche of further questions, as the group realised how shaky their grasp was of the basic ideas – What is pressure? What is air? Why do your ears go pop? What about people living high up in the mountains? A great opportunity to explore the nature of the air's invisible presence.

CABIN PRESSURE

It's true, the pressure of the air inside a plane has to be kept up artificially as a plane ascends because the pressure of the air outside lessens with height. This is achieved by 'bleeding' off air that has already been compressed inside the jet engines. If this were not done, air pressure would reduce dramatically inside the plane just as it does when you climb a high mountain. This would leave fewer and fewer oxygen molecules for us to take in with every breath. We'd soon get light-headed, as can happen when you are high up a mountainside, and ultimately black out.

But the fact that our ears pop as a plane ascends and descends shows something more: although the pressure inside a plane is kept artificially higher than the pressure outside, it is still a bit lower than it had been at ground level. This precaution is taken to ensure the difference in pressure between the inside and outside is not too extreme; otherwise there'd be extreme stresses on the structure of the plane and a risk that it could explode. The cabin pressure is maintained at roughly the same level as at the top of a mountain – roughly 2,400 m, the height of Mexico City. Our

DOI: 10.1201/9781003272779-14

ears 'pop' whenever there is a significant difference between the pressure inside the ear and outside it; reduced pressure inside a plane is enough to make this happen, just as it does walking up a mountainside. 'Popping' helps equalise the pressure across the ear drum.

THE EAR

The fact that there's a 'drum' in the ear is pretty well known; but what does this mean? The eardrum is a membrane in the ear canal that separates the external ear from the middle ear. Physically, its job is much like that of a musical drumskin: to vibrate. But whereas a musician's drum is hit by a stick to get it vibrating, the eardrum is set in motion by tiny variations in the pressure of the air surrounding it. These vibrations are what a sound wave is: tiny fluctuations sent out from the voices, music and noise that surround us, as the elegant engraving in Figure 13.1 illustrates.

It's hard to believe that such delicate, imperceptible vibrations that you can hardly feel nevertheless get picked up by the eardrum. It is indeed remarkable: the minute changes in air pressure when a typical sound wave passes are around 2 units (called Pascals) on top of the normal atmospheric pressure of 100,000 units. In other words, the eardrum responds to pressure fluctuations of just 1 part in 50,000 or 0.002%. That's how sensitive it is.

What about the other side of the eardrum? What is it that gets blocked when you have a cold? The middle ear is a chamber containing three miniscule bones (called ossicles) linked to one another (Figure 13.2). The first of these is physically attached to the eardrum and together the three of them amplify and transmit vibrations from the outer ear, via the eardrum to the inner ear, where the cochlea is located. It's in the

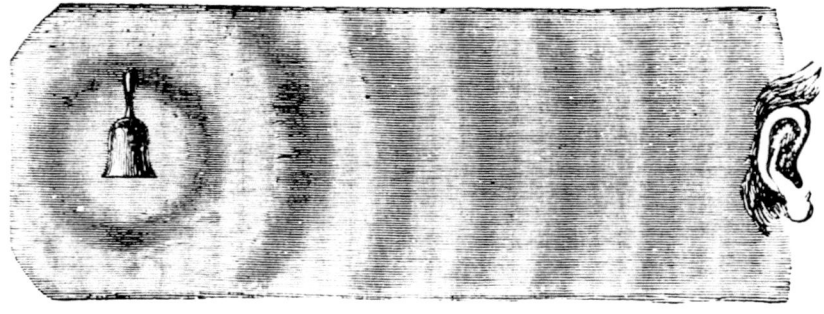

Figure 13.1 Engraving of sound waves travelling to the ear.
(From Popular Science Monthly 1878, via Wikimedia, Public Domain.)

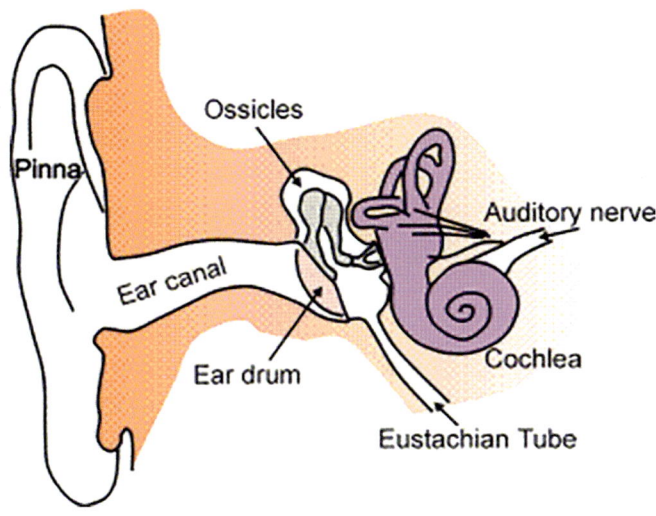

Figure 13.2 Parts of the human ear.
(Image credit: Iain at English Wikipedia, https://creativecommons.org/licenses/by-sa/3.0/deed.en)

liquid-filled inner ear that the vibrations stimulate nerve signals. These pass along the auditory nerve to the brain, where they give rise to the perception of sound.

Now back to the blocked ear... the middle ear is a chamber ordinarily filled with air at the same pressure as the outside atmosphere. If the external pressure were to go down, as it does when an aeroplane climbs high, an imbalance is created with the pre-existing pressure in the middle ear. This gets rectified thanks to a small tube (Eustachian tube) which links the middle ear to the outside world. Normally this is closed, isolating the middle ear from the air outside. But when there is a pressure difference across the ear drum, it opens up, connecting the inner ear, via the nose and throat, to the atmosphere outside. A bit of air rushes in or out to equalise pressures – a great mechanism for protecting your ear. This is what happens if you yawn or move your jaw when you enter a tunnel in a fast train – you are opening up the Eustachian tube temporarily.

However, this tube can get blocked when you have a cold. It can get inflamed or filled with mucus and fail to open. You can see this explained visually in a helpful short video at https://www.youtube.com/watch?v=XctwiScoorM. So, somehow, Julie's blocked-up Eustachian tube must have temporarily opened as the pressure inside her plane reduced from its ground level value, thus relieving her ear drum and restoring normal hearing.

As often happens when exploring a real-life event in a discussion group, questioning moved on rapidly: in this case from the biology of the ear and the physics of sound, to a more basic issue: What actually is air pressure? Why does it vary? and, as Marian queried: 'What is it that makes the weight of air. Is it the water in it?'

AIR

Air fascinates us! As mentioned in Chapter 5, discovering that air is a physical substance – a mixture of various gases – is an important part of our early science education. Perhaps the most difficult thing for us to accept, even as adults, is that air actually has weight. Grappling with this issue in the group, Sonya asked, 'if you take a jar of air and a jar of vacuum do they weigh differently?' Good question – yes they do, but of course, a jar of air weighs so little, the difference would be undetectable with kitchen scales. Finding a jar without air in it, for comparison, would also be quite a challenge!

THE PRESSURE OF THE ATMOSPHERE

Sitting here on the surface of the Earth, however, places us underneath a huge amount of air: there's about 100 km up to the conventional 'boundary' between the Earth's atmosphere and the outer space (though there is no boundary in reality – just a gradual thinning out). Even with its very low density, the great column of air above our heads weighs a lot because it is so tall: about 10,000 kg of it sitting above every square metre. That means directly above our bodies (approximately 1/10th of a square metre) is about a tonne of air! That's why the air around us is pressurised – it's the weight of the air above it pressing down. Fortunately for our wellbeing, the air inside and outside our bodies is connected, so both are at the same pressure. This explains why our lungs and inner ears don't explode under the pressure of the atmosphere.

A simple way of showing how the pressure in a fluid varies with depth is given in Figure 13.3: a photograph of a pierced bottle of water. The pressure at any given level is simply caused by the weight of the fluid above the level. The water at the bottom sits under the greatest weight of water above, so is more pressurised and spurts out faster. The same goes for the air in the atmosphere above us. The weight of all the molecules in the air – nitrogen, oxygen and some water vapour too – is pressing down.

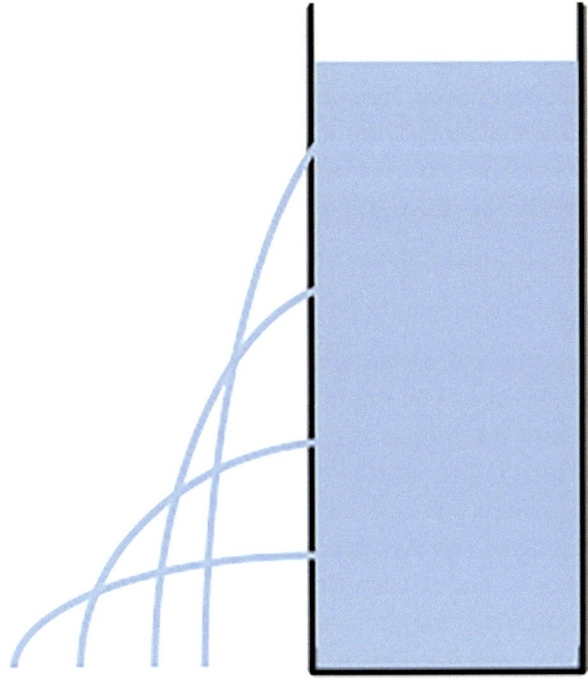

Figure 13.3 Pierced water tank showing greater pressure at lower depths. (Image credit: Mirjam Glessner.)

ULTIMATE CAUSE OF PRESSURE

At the microscopic level, any gas is made of countless billions of molecules rushing around at very high speed. Pressure is simply the net effect of vast numbers of these molecules hitting any given surface. They exert a force on a surface, as a jet of water from a hose does against a wall. The molecules rushing around in the atmosphere hit your body, and, in particular, the sensitive part which is your eardrum. They cause no damage under normal conditions because molecules in the inner ear push back with an equal force.

You can see how the idea of molecules rushing around in a gas helps explain simple things we are familiar with. For example: when you pump up tyres, you are simply pushing more molecules from the pump into the inner tube. So, more molecules are hitting the walls of the inner tube, increasing the force on it, thus raising the pressure. Another way of increasing pressure is to reduce the amount of space for the molecules to move around in so they hit the walls more often, again increasing the average force on the walls and hence the pressure. This is what happens when you squeeze a balloon. Yet another way to raise the pressure of a gas is to raise its temperature. This makes the molecules rush around even faster and so hit the walls with even greater force. That's why you shouldn't leave an aerosol container lying in the sun – it increases the pressure in a confined space which could cause it to explode. It's also why it's wise to measure the pressure of tyre before driving off, while it's still cold.

THE NATURE OF A GAS

The original question sparking off this discussion was inspired by a plane journey and a head cold. It led, as you might expect, into an exploration of air pressure, sound and the human ear. More surprising, however, was the turn taken next. Trying to visualise a gas, not as some kind of continuous, nebulous medium, but instead as an enormous collection of tiny molecules, Mary asked somewhat profoundly: 'What is it that's in between the molecules in air, or any other gas?' Simple though the question was, the answer proved hard to accept. 'Is there really nothing in between the molecules?' was Sonya's quizzical response. 'What do we mean by nothing, anyway' she added. It was a similar deep question to that raised in Chapter 7, in relation to the structure of ice.

This is indeed how we imagine any space not occupied by a particle: it is simply part of the all-pervasive vacuum. Air, or any other gas, is not a continuous medium, but mainly a huge emptiness with specks of matter (molecules) dispersed sparsely in it. The vast reaches of outer space are devoid of matter (almost) and so too are the inner spaces of gases such as air. Grappling with the implications of this emptiness taxes philosophers, mystics and, more recently, quantum physicists to this day.

This picture of the microscopic nature of a gas – unimaginable numbers of tiny molecules buzzing around at great speed in a vacuum – is one of the fundamental concepts in science. It helps us understand many important processes in modern life: compressing oxygen for use in hospitals; designing efficient aircraft wings to minimise fuel use and, of supreme importance today, countering changes in the Earth's atmosphere caused by rising levels of carbon dioxide. One more example of how understanding of key concepts in science can help us all make sense of the world in which we live.

Fourteen

Worrying cough or just a tickly throat? Another pizza for the kids or tantrums over salad? Heat pump now or wait till prices drop? Tough decisions confront us every day yet, all too often, the evidence we need to make a choice is inconclusive or missing. The Covid-19 pandemic, arising unexpectedly and progressing unpredictably, has demonstrated starkly how decisions sometimes have to be made even when evidence is lacking. The accelerating climate emergency, on the other hand, shows how evidence, even when it's strong and plentiful, may still not sway us. Threats to our health and that of the planet are forcing difficult choices upon us with increasing urgency today, but decision-making has, of course, been an all too familiar feature of everyday life since time immemorial. Fortunately, it's also a major focus for scientific research in the fields of experimental psychology and neuroscience. In this chapter, we focus on what we can learn from scientific studies of the everyday act of choosing.

DECISION-MAKING

Philosophers have long pondered the way we make decisions. In the seventeenth century, the idea developed that we should weigh up the pros and cons of each possibility and then factor in its likelihood. It's a bit of a pain carrying an umbrella to work but if the chance of rain is high, you take one; if it's low, you don't bother. More sophisticated theories have developed over time to take account of other factors, such as how well you personally cope with risk and what your individual preferences are. Classical economic theory assumed that, as consumers, we make rational choices in our best interest. It's really quite recently that research on our decision-making behaviour has moved onto a more realistic level, through the use of experiments. Experimental psychology researchers set up games or situations that simulate real life events and study how people react. This kind of research has thrown up a number

DOI: 10.1201/9781003272779-15

of interesting findings, that are sometimes firm enough to be replicable whenever they are done – a hallmark of sound science.

HEURISTICS

The Nobel Prize was awarded to Daniel Kahneman for work of this kind… and much more. With Amo Tversky, he developed a theory that combined insights from both psychology and economics. They identified ways in which we cope with decision-making by using heuristics and biases to simplify the process. Heuristics are simple strategies or mental processes we use to reach decisions or make judgements … known colloquially as rules of thumb.

Various kinds of heuristic have been identified by psychologists, many of which we may recognise in our personal lives. If we are trying to work out the rough price of something or the area of wall we need to paint, we may round up the numbers to make calculation easy – 6 bread rolls at 48 p each is roughly 6 × 50 p or £3: a rule of thumb. An educated guess is another example in which you use past experience to make a reasonable guess without investing too much time and effort in researching carefully. Another well-known shortcut is believing someone just because they are in authority – the power of the white coat (see Figure 14.1) – as in the example of the Milgram experiment described below.

BIASES

Heuristics are clearly absolutely essential ways of getting through the complexities of real life without collapsing from mental exhaustion. However, they can lead all too easily to bad judgements and decision-making.

Psychologists have studied these tendencies in experiments that simulate real situations. They find various ways in which errors, biases and deceptions creep in. Figure 14.2 illustrates a classic type of bias, resulting from having survived a challenge.

Another type involves registering a piece of information too strongly just because it came first. For example, a salesperson may show you an expensive item first, so when you are shown something less expensive it seems more affordable. Another way for error to creep in is by overestimating whether something is likely to occur just because we remember cases that stick in our mind – like plane crashes or shark attacks.

We are also inclined to judge risks according to our feelings, even if this goes against the facts. An interesting study, by Paul Slovic in 2012, of

Figure 14.1 The power of the white coat.
(Image credit: Dan Piraro.)

the risk from harmful radiation showed that we, the general public, rate the risk from nuclear power and waste as very high and that from medical X-rays as low, whereas the majority of radiation experts see it the other way round. A consequence of this bias is that the cost of interventions aimed at saving lives may be out of proportion to the benefits. For example, one study estimates that the average cost of saving one year of life in the nuclear industry is around $100 million, whereas by making seat belts compulsory an equivalent saving of lives was estimated to have cost a mere $69.

A further way in which our feelings may introduce bias was discovered through experiments that use small amounts of money to simulate real life transactions. These suggest that, when weighing up options, we tend to be more negatively affected by losing something than we are positively affected by gaining something equivalent.

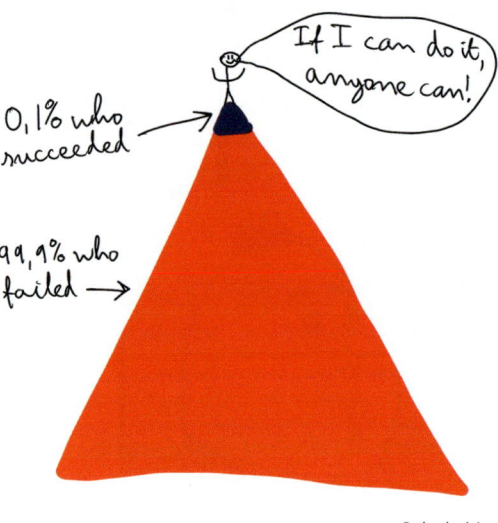

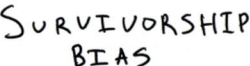

Figure 14.2 A classic type of bias.
(Image credit: thedecisionlab.com.)

FRAMING

A different kind of influence on our decision-making comes from the way things are framed for us. In a classic experiment, Kahneman and Tversky organised two groups of people in an experiment about taking a medical decision. The first group was asked to choose between two treatments for a deadly disease. One treatment would save 200 lives out of 600; the other offered a 1 in 3 chance that everyone would be saved but a 2 in 3 chance none would be. Most people decided that the 200 people should be saved – the less risky option. A different group of people were told the same information but framed differently – *losing 400 lives rather than saving 200 and a 2/3 chance that all would die.* With this more negative framing, the majority of this group chose to take the high risk, with the slender chance of preventing all the deaths.

The upshot of this kind of research and the kind of behavioural science that is developing from it is that it can't be right to assume we all act rationally when we make decisions. We have evolved to use shortcuts that introduce bias, error and deception…. but, on balance, that's better than pure guesswork!

CHOOSING WISELY

Once we become aware of quirks in the way we make decisions, we are better placed to make good decisions. Research undertaken in 2006 by Ap Dijksterhuis and colleagues in Amsterdam showed that sometimes making a snap judgement can be better than deliberating carefully. Too much information can be a problem in situations like picking a holiday destination or dishes from a menu. The researchers concluded that it's better to think things through when choosing is relatively simple, but for complex decision-making, less analytical thinking may be preferable.

Any of us may have witnessed a situation where someone persists with a choice even after they know it's not right – they're so far invested in it. Studies at Ohio State University indicate that this happens because we feel more committed the more we invest. Sometimes, it's better to just let the past go – to stop throwing good money after bad.

Many studies reveal that almost all of us are susceptible to peer pressure. The famous Milgram study at Yale University in the 1960s caused shock waves by demonstrating how far volunteers were prepared to go in administering electric shocks to a (stooge) victim if a figure of authority in a white coat told them to do so. It is believed that a fatal plane crash in the East Midlands in the UK occurred after a member of the cabin crew, though realising the captain had shut down the wrong engine, decided not to question his authority. Being aware of the pitfalls of peer pressure helps with better decision-making.

A further challenge in today's consumer society can be an excess of choices. It's easy to be overwhelmed by the sheer number and range of differing options – particularly in a modern coffee shop (Figure 14.3). This problem was investigated by Barry Schwartz, a psychologist in Pennsylvania, who studied the choices faced by students looking for jobs on leaving university. It turned out, as might be expected, that those who looked thoroughly at every choice were more likely to move on to higher salaried posts than those who had just adopted a 'good enough' approach. More surprisingly, however, it was found that these same higher earners also tended to be less satisfied with the choice they had made, compared to those who had been more easy going about choosing. Some reassurance for those who make their choices less earnestly!

RESISTING EVIDENCE

Ignoring or avoiding evidence, whether scientific, legal or personal, is, unfortunately, a rather common aspect of everyday life. We're all

Figure 14.3 Sign outside coffee shop.

reluctant to cut back on pleasures even when the evidence of their damaging effects is plain to see. It's not easy to give up smoking, or to cut back on red meat or holiday flights however aware we are of evidence on the harm they might cause. It's even more tricky when the evidence runs against our personal or commercial interests.

Research in psychology reveals a further basis for resistance: cognitive dissonance – the theory that people try to avoid inconsistencies in their understanding or beliefs. This avoidance reaction can lead us to ignore unwelcome information, deny it or shut it away in a discrete part of our minds ('compartmentalising' it).

A classic example of professional denial is the case of handwashing. Ignaz Semmelweis was laughed at and ignored by his colleagues when he argued in 1850 that washing hands by medical staff could lower the rate of hospital-acquired infections. The theory that diseases could be spread by germs had not yet been developed and some doctors felt professionally offended by the idea that their hands were unclean. What seems common sense to us today ran counter to prevailing ideas about disease at the time.

Perhaps the most egregious example of denial for commercial reasons in recent times was the effort made by the tobacco industry to cover up harm from smoking. It commissioned pieces of so-called 'independent' research with the aim of creating conflict and sowing confusion. Widespread public denial of fundamental scientific evidence was revealed by a Gallup survey as recently as 2017. It found that about 38% of adults in the United States deny Darwinian evolutionary theory believing that 'God created humans in their present form, at one time within the last ten thousand years'. A 2015 study by Dan Kahan at Yale University found that such denial was not associated mainly with ignorance of the scientific facts but with political and ideological conservatism (Figure 14.4).

Clearly, the consequences of rejecting scientific evidence about handwashing, smoking or climate change are very serious. They put the health of individuals, communities and indeed the planet in jeopardy. Yet avoiding unwelcome evidence is a normal aspect of being human.

Figure 14.4 Alternative evidence.
(Image credit: James MacLeod macleodcartoons@gmail.com.)

There are situations in which it may even be to some extent necessary. For example, a study in the field of nursing found that in some cases self-deception can help protect a vulnerable patient. Some distortion of reality can have a positive effect by enhancing the sense of control, lowering levels of anxiety and helping with decision-making under stress. As the researchers put it: the benefit of seeing the world through rose-tinted spectacles.

PATCHY AND DODGY EVIDENCE

Even experts can run into difficulties when evidence is patchy. In such circumstances, they tend to move from the established facts to exercising their judgement, according to a 2019 article in *Scientific American* by Baruch Fischhoff, a specialist in risk analysis. In some circumstances, this may be good enough: forecasting tomorrow's weather is an example. The judgement made one day is tested by reality the next. In uncharted territory, however, when feedback from the real world is not available so immediately, experts may well be in much the same position as the rest of us. The ups and downs in the early stages of the COVID-19 pandemic revealed this, as the virus mutated unpredictably and patterns of social behaviour emerged in unforeseen ways.

Worse than patchy evidence is deliberate misinformation. Clearly, this is regularly put about by individuals, and organisations and its reach is being dramatically extended, thanks to social media. Ethically motivated organisations and individuals now have to work even harder to ensure that their communications are as effective and far reaching as possible. An example of this was the Symptom Tracker study from Zoe and Kings College London which attracted over 4 million participants who reported daily on their symptoms and experiences during the pandemic. One of its studies looked at the reasons people give for their reluctance to accept vaccination. It found these to be overwhelmingly linked to ignorance of the facts and fears about possible consequences; they were only marginally associated with people's ideological or religious beliefs. Spreading of myths and false information through social media may well have contributed to vaccine hesitancy.

CONCLUSION

This small sample of psychological and behavioural studies about the ways in which we make decisions confirms that we are not simply the rational actors assumed by classical economics; we are liable to bias, prejudice and denial. In many ways, these tendencies are essential for our survival, bombarded as we are by excess stimuli and pressures. On the other hand, they can cloud our reason and lead us to make poor choices and damaging decisions. Science itself aims to minimise the effects of bias and prejudgement. The insights it offers into the decision-making process may not always save us from ourselves, but they can at least put us on the lookout for our errors.... And help us make allowances for them.

Fifteen

A pandemic on an unprecedented scale in modern times continues to devastate the world as I write. Who knows how it will be at the time you are reading this. What we know about the behaviour of the virus and the pattern of the epidemic is steadily shifting. Research evidence inevitably lags behind events under these conditions; websites and news bulletins, rather than books and journals, are needed to keep up us up to date. This chapter is no substitute for that: its purpose is to explain some of the key scientific concepts, so that emerging news is less baffling. So many issues are raised – in biology, pharmacology, social behaviour, mathematics and a host of other disciplines – that two linked chapters are devoted to this topic. We start with viruses, lungs and epidemics. In the following chapter, we look at immunity, vaccines and variants.

VIRUSES

The image in Figure 15.1 is a representation of a few of the coronaviruses known officially as SARS-CoV-2, generated by a computer from a photograph taken under an electron microscope – a very high magnification microscope. It shows that the outer surface of the virus is spherical and studded with proteins. This virus caused the outbreak of respiratory illness first detected in Wuhan, China in 2019. The illness caused by the virus is named after that fateful year: COVID-19.

The image in Figure 15.2 was made from a patient in the USA. It was taken through an extremely high-powered microscope, known as a scanning electron microscope, and shows multiple copies of the virus (the round, gold-coloured objects) emerging from the surface of cells cultured in the lab. It gives a sense of the relative sizes: viruses being much smaller than cells; the coronavirus is roughly 100 times shorter in length than a cell.

Strange though it may seem, it's an open question whether a virus is a living thing or not. It depends on your definition of 'living'. Virologists describe them as leading 'a kind of borrowed life'. This ambiguity arises

DOI: 10.1201/9781003272779-16

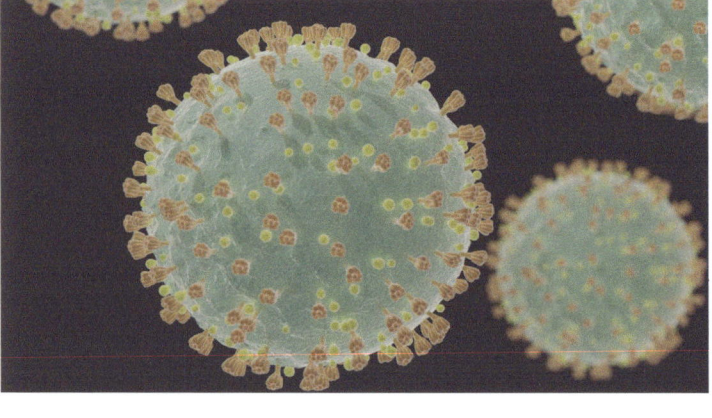

Figure 15.1 Coronavirus (SARS-CoV-2) through an electron microscope. (Image credit: Felipe Esquivel Reed, https://creativecommons. org/licenses/by-sa/4.0/deed.en)

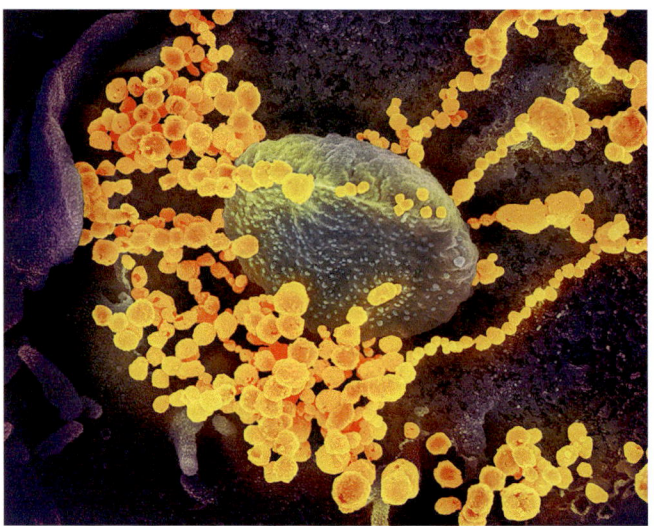

Figure 15.2 SARS-CoV-2 viruses emerging from a cell. (Image credit: National Institute of Allergy and Infectious Diseases (NIAID).)

because they are unable to reproduce unaided; they rely on breaking into living cells and commandeering their machinery for this purpose. This is why they are so dangerous: the cells they enter are destroyed in the process.

The model of a coronavirus (not specifically the one causing COVID-19) in Figure 15.3 shows its internal structure.

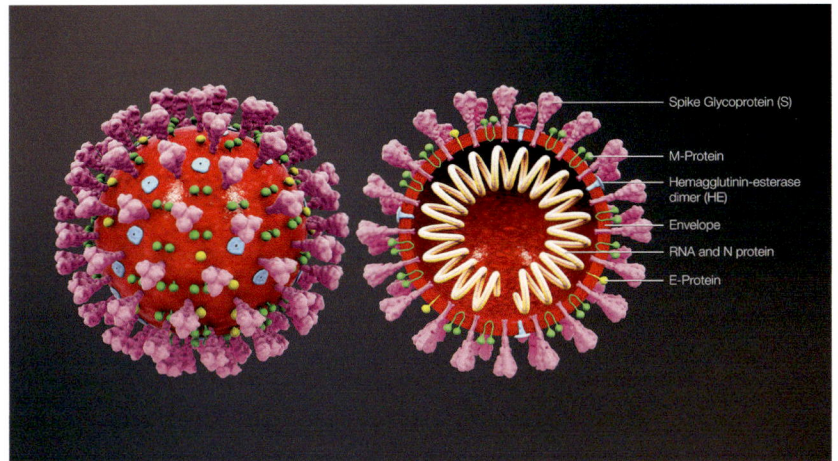

Figure 15.3 Model of a coronavirus (similar to the one causing COVID-19). (Image credit: scientific animations via Wikimedia, https:// creativecommons.org/licenses/by-sa/4.0/deed.en)

It's an amazing assembly of large molecules – an extraordinary piece of engineering. On the left, you can see its outer shape – a sphere in this case, but quite different in other kinds of virus. The red coloured surface (envelope) is made of oily molecules (lipids) and is studded with various types of protein molecule that penetrate through this thin shell of lipid molecules into the interior of the virus particle. Following fast moving research by a team at the University of Texas, the so-called 'spike' protein was identified in February 2020 as the key molecule that locks onto cells in the lungs. It's the big pink shape in the diagram, labelled 'spike glycoprotein S' (a glycoprotein is a protein molecule that has some carbohydrate molecules attached).

The envelope is a protective container inside which lies the blueprint for making more viruses – the RNA. This is a very similar molecule to DNA, but is made of a single rather than a double strand. It serves the same function, providing the vital genetic information for the production of new viruses. Unlike living cells, however, a virus does not contain the machinery needed to use this blueprint to reproduce itself. Instead, it breaks through the outer surface of a living cell, much larger than itself, and uses the cell's machinery to reproduce its RNA in place of the cell's own DNA. It's a freeloader!

SARS-CoV-2 is a specific, recently mutated version of a coronavirus. The coronavirus family is very familiar to virologists as various of its members have been known about since the 1930s, though it was only recognised as a family and given its name in the 1960s. Various kinds of coronavirus are associated with the common cold and more severe diseases such as Middle East Respiratory Syndrome (MERS) and Severe Acute Respiratory Syndrome (SARS).

LUNGS

The coronavirus infects cells in the respiratory tract, which runs from the nose to the lungs. The virus latches onto receptor molecules which lie on the surfaces of specific types of cell. Receptors are large protein molecules, lodged in the membranes of cells, to which other molecules can attach.

Figure 15.4 shows the airways of the lung branching out across the lungs from the central windpipe (trachea). The tip of one of the smallest branches is enlarged to show how the cavities (pinkish balloon shapes called alveoli) that contain the air you breathe in are surrounded by tiny blood vessels. This is where oxygen passes through the very thin tissue of the airways into the arterioles (red) and carbon dioxide passes out the other way, from the venules (blue) into the airways of the lung, ready to be expelled.

The cells that line this cavity are one of the targets for the coronavirus. Like the virus itself, these cells also have protein molecules projecting out

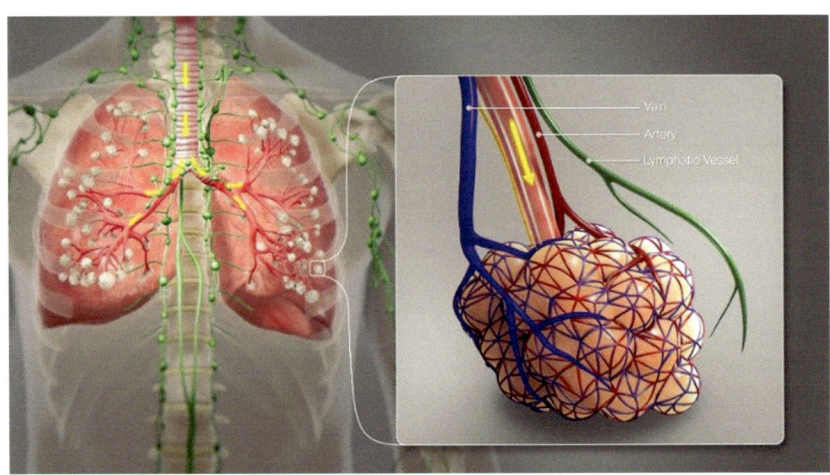

Figure 15.4 Model of the lungs with expanded view of alveoli.
(Image credit: Scientific animations via Wikimedia, https:// creativecommons.org/licenses/by-sa/4.0/deed.en)

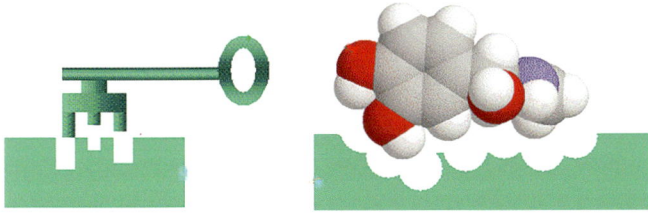

Figure 15.5 'Lock and key' model of a molecule interacting with a receptor. (Image credit: Brookhaven National Laboratory.)

of their surface. The crucial interaction occurs when the protein sticking out of the virus 'matches' the protein sticking out of the cell. 'Matching' here means that the protein on the virus has complementary features to those on the receptor on the surface of the cell, rather as a key has a complementary shape to the lock into which it fits. This concept was previously encountered in Chapter 9 in relation to enzymes (Figure 15.5).

In March 2020, just a few months after the outbreak of COVID-19, a team at Westlake University in Hangzhou, China identified the structure of the receptor protein on the surfaces of the respiratory cells and worked out how it interacts with the protein on the surface of the virus. Knowing the structure of the key (the virus protein) and the lock (the receptor protein in the lung cells) could in principle be a starting point for developing drugs to treat the disease. In general, pharmacological research proceeds on two distinct fronts when confronting a new viral infection: firstly, looking for a vaccine, and secondly, looking for an antiviral drug. The 'spike' protein on the surface of the virus was identified early on as a promising molecule from which to prepare a vaccine. It turned out that this approach worked and vaccines were developed rapidly over the following year. Their purpose was to stimulate the immune system to destroy viruses emerging from infected cells before they were able to multiply further.

A well-made animated video explains clearly the whole process of the disease, its spread and how we can slow it. It's available at https://uncrate.com/video/coronavirus-explained/?fbclid=IwAR2FRflUuMKsWeMcOC_7db08bfNLuNhGkIr8XSEDfwTl6W6IJJU7AZGkIZA

EPIDEMICS

Our immune systems are coping continuously with threats that surround us on a daily basis. They generally succeed – that's how we stay healthy.

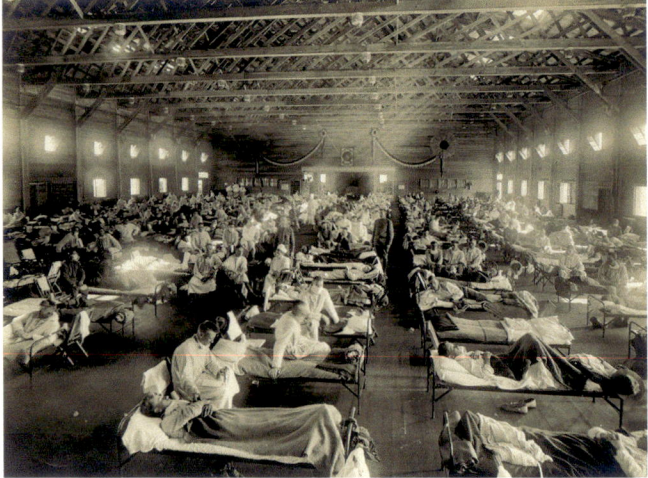

Figure 15.6 Emergency hospital during the 1918 influenza epidemic. (Image credit: National Museum of Health and Medicine via Wikimedia, Public Domain.)

Since birth, cells in our immune system have been building up a wonderful repertoire of different antibody molecules, ready to fend off microbes. But, as we all know only too well, this doesn't always work. We still get colds and flu from time to time. Bugs can be 'passed around', but in normal circumstances, they eventually die out. A bout of colds arises when, on average, one person infects more than one other person: the bug spreads. They die out when an infected person infects less than one other, on average over a given population. Serious epidemics occur when the infectious agent passes easily and rapidly from person to person. The degree of seriousness depends on many factors including the length of the incubation period, the virulence of the bug, the geography of the population and its social behaviour.

The photo in Figure 15.6 of an emergency hospital in Kansas during the 1918 influenza epidemic (*courtesy of Wikimedia*) reminds us that epidemics are not just a creature of globalisation but have occurred throughout history – from Athens (430 B.C.) via the Black Death (1340s/50s) to HIV, Ebola and Zika in more recent times.

EXPONENTIAL GROWTH

We are accustomed to all kinds of changes that happen over time: the rise and fall of daylight, the passing of the seasons, the slow progress of a ship traversing the oceans, for example. Such changes tend to be relatively steady, with roughly equal differences occurring in equal intervals of time. A ship covers approximately the same distance each day. Some kinds of changes are, however, far more dramatic. A popular tweet might spread to ten followers, each of whom re-tweet to ten of theirs, reaching a hundred more. Then a thousand, ten thousand and so on. The message will have 'gone viral'. In successive intervals of time, the number of retweets is not the same; instead of say ten retweets being added every second, the number multiplies. This kind of growth is called exponential. The change *multiplies* in equal intervals of time rather than adds. Instead of a steady addition from, say, 10 to 20 to 30 to 40, you have multiplication form 10 to 100 to 1,000 to 10,000….. or, to put it more neatly, 10^1 10^2 10^3 10^4. The small number (the superscript), known as an 'exponent', lends its name to this kind of growth: exponential. It's easy to see how this pattern of growth characterises the spread of a pandemic. If a carrier infects two people and they infect two more and so on, the total number multiplies exponentially, as shown in the upper left-hand side of Figure 15.7. If instead each person was to infect five, the growth would, of course, be far more rapid.

In the case of COVID-19, the average number of people an infected person infects (known as the reproduction number) has varied over time depending on the measures taken to curb its spread and on the emergence

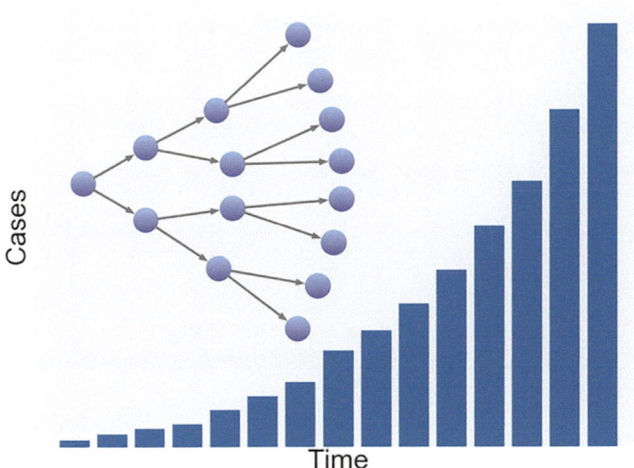

Figure 15.7 Exponential growth with a reproduction number of 2. (Image credit: Dennis Liu, permission granted.)

of new variants. The reproduction number of the original coronavirus first detected in Wuhan was estimated to be between 2 and 3. The omicron variant spreading rapidly at the time of writing (December 2021) is estimated to be between 3 and 5. For seasonal flu, the figure is lower than for either variant.

The exponential growth phase of epidemics doesn't last forever. In a finite population, the rate of growth of disease must eventually stop increasing as fewer and fewer people remain to be infected. The daily growth in numbers ceases to be exponential; it gradually diminishes from day to day, tending towards a steadier daily figure (the 'plateau') that doesn't grow from day to day. Eventually, the daily figure begins to fall, reaching a low or manageable number. The graph in Figure 15.8 exemplifies these stages in the case of flu. Using consultation rates for local doctors over the course of the winter in four separate years, the four curves show how numbers rise gently at first, then steeply until they peak or flatten off dying away ultimately in reverse fashion. The graphs also show how the extent of the infection varies from year to year, only reaching epidemic proportions occasionally (as in the red curve). Diseases are not usually eliminated entirely from a population.

To combat the effects of an epidemic, several strategies are available. The immediate spread of the pathogen can be slowed down by minimising the degree of contact between people, so those who carry the bug are less likely to pass it on to others. As became all too familiar from the early

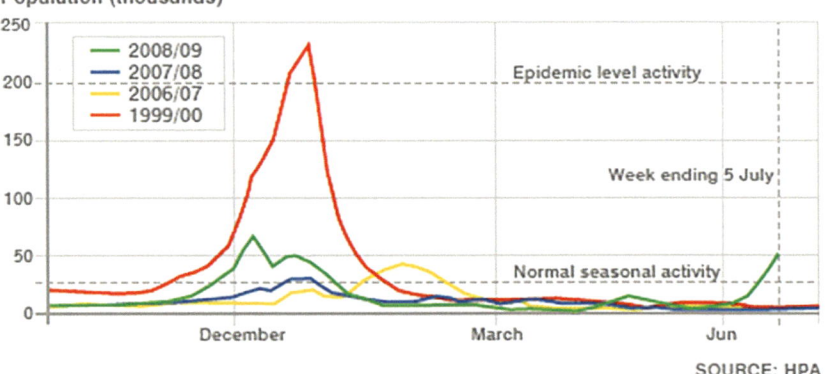

Figure 15.8 Rates of flu infection in four specific years.
(Image credit: UK Health Protection Agency, Open Government Licence v3.0.)

days of the COVID-19 pandemic, this led to changes in behaviour, either on a voluntary basis or mandated by Governments. Strategies included guidance to individuals to maintain a distance between themselves and others; a requirement to wear face coverings in crowded situations and advice or legal requirement for people to work from home wherever possible. The effectiveness of these measures depends to some extent on the way messages based on complex scientific evidence are formulated and communicated to the general public.

INFLUENCING SOCIAL BEHAVIOUR

The course of an epidemic is critically dependent on changes in people's everyday behaviour in simple but important ways. Social, economic and psychological scientists carry out research through surveys, experiments and studies of individuals, groups and communities about many aspects of behaviour. Public messages about hand washing and social distancing, for example, are not just dreamed up but are based on rigorous studies.

The example of hand washing, however, shows there's often a gap between scientifically based recommendations and actual behaviour. Research by Burton and colleagues at the London School of Hygiene and Tropical Medicine in 2011 shows that washing your hands with water alone reduces the quantity of germs by 50% – and adding soap reduces it by 80%. Yet, in reality, people report that they wash their hands much more often than they actually do. They also wash less when they think they are not being observed. Studies of hand washing behaviour suggest that some people may not bother because they are simply not aware of how well specific handwashing actions can prevent the spread of infectious disease. Others may feel it's less effective or novel than wearing a mask, and so discount it. Simply making water, soap and drying materials more available in public places is the kind of simple measure that has been shown to have a positive effect on behaviour.

A Behavioural Insights Team, originally set up by the UK Government, has used behavioural research to establish a set of principles for government communications. These include maintaining public trust; making messages clear, simple and precise; and being transparent. The team recommends that instructions are more likely to be followed if they are clear and easy to remember. They also suggest that showing what the authorities are doing behind the scenes may help improve public perceptions of those actions.

ANTIVIRAL DRUGS

A second strategy for defeating a novel disease is to invest resources in developing drugs to combat it. One approach, of particular importance in the immediate circumstances of a novel epidemic, has been to see whether existing treatments can alleviate some of the most threatening effects of the disease. Examples of this include anti-inflammatory drugs to reduce inflammation in the lungs and therapy using blood plasma from recovered patients to boost the immune system.

A more directed approach is to try to find a drug that directly tackles the virus itself, as an antibiotic does for bacteria. Potential drugs could be aimed at interfering with different stages of the virus life-cycle: preventing a virus entering a cell, for example, or stopping it reproducing if it does so. Different research teams work on different approaches. The challenge is, however, huge: 'new territory', in the words of a virologist at Loyola University in Chicago. Antiviral drugs, in general, are particularly hard to develop because of the risk of damaging the very cells the virus is entering. Some have nevertheless been successfully developed and are on the market. A drug called aciclovir (Zovirax), for example, attacks the herpes virus responsible for cold sores by damaging its DNA without affecting that of the host cell.

One promising route, particularly for responding rapidly to an emerging threat, is to see whether drugs already proven to be effective against a different virus might also work against the COVID-19 virus. The advantage of 're-purposing' existing drugs is that they will have already been tested for safety, toxicity and ability to remain intact in the body. This shortens the time needed for trials, though is something of a 'hit-and-miss' approach. In March 2020, a drug, known as Favilavir or Avigan, was approved in China for the treatment of influenza and also for use in clinical trials for COVID-19. The drug interferes with a crucial enzyme that enables flu viruses to multiply. Fortunately, coronaviruses rely on the same enzyme, so the hope is that it will also prove effective in the treatment of COVID-19. Clinical trials were underway at the time of writing (Winter 2021).

VACCINES

A third strategy for combating infection is to seek to develop a vaccine. Vaccines employ a completely different method to combat infection. Rather than directly interfering with the virus itself, they stimulate the body to tackle the pathogen through its immune system.

A vaccine works by provoking the immune system to produce anti-body molecules and defensive cells that help the body fight off unwanted pathogens, such as viruses, bacteria and fungi. One type of vaccine consists of some disabled fragment of the offending molecule (known as an antigen) – for example, it may just be part of one of the protein molecules on the outer surface of a virus. This can be sufficient to trick the immune system into developing antibodies and defensive cells that destroy the virus. Antibodies are giant protein molecules, produced in specialised cells of the immune system, which attach to the alien molecules and signal to the body's defences that they are to be disposed of. Crucially, the immune system not only fights off unwanted pathogens such as viruses and bacteria but also keeps copies of the antibodies that achieved this, effectively providing a 'memory' of an unwanted pathogen ready to mount a renewed defence were the virus to present itself months or years later.

Scientists across the world worked overtime in 2020 to find some kind of fragment or disabled version of the 'spike' protein on the surface of the virus that would provoke the body to produce an immune response. Developing a practicable vaccine is no easy task: it not only has to be shown to be effective for a large enough proportion of those vaccinated, it has to be proven to be safe as well. As it turned out, the short period of time it took to develop and test the first vaccines against COVID-19 was an extraordinary and unprecedented feat of research and development.

The immune system and the vaccines aimed at stimulating it to fight off the coronavirus are of such central importance in the COVID-19 story that the next chapter is devoted to exploring them in greater detail.

More about the way drugs and vaccines are developed can be found in a blog on the website: 2.10 Drug and vaccine development | Getting to Grips with Science (gtgwithscience.com) (https://gtgwithscience.com/blog-series/science-blogs/our-bodies/2-10-drug-and-vaccine-development/).

Sixteen

As the pandemic swept across the globe, immunity became the great hope of humanity. This wondrous condition promised to free us from the worst excesses of the virus and pave the way to our former ways of life. But, as we were soon to learn, immunity is not a permanent condition; its strength can wane with time. Nor is its protection guaranteed: reinfection does occur. Questions began to be asked: What exactly it is? How do we attain it? How long might it last? Will it survive the onslaught of the variants?

With change happening at such a pace, information at the time of writing (Autumn 2021) may be out of date by the time you read it. Scientific research can only track and model the evolution of the virus. These urgent questions can only be answered in real time as studies assess the effects of the unfolding disease. It's through the daily news bulletins, not the textbook, that we're learning much of our science now. Here, we explore what is known about immunity and its limitations, to help us interpret the news that's yet to come.

IMMUNE SYSTEM

It's a *system* – that's the first striking feature of our defence apparatus. It's not a discrete organ or chemical that fights infection. It's a system that has evolved gradually over the aeons, as simple organisms gave way, first to sea creatures, then their counterparts on land and ultimately higher mammals. Even simple sea sponges have some (unconscious) sense of the distinction between self and other: they are able to recognise tissue that has been grafted as not their own. Higher up the evolutionary tree, we find a more sophisticated apparatus for dealing with alien material: specialised cells that are capable of engulfing and destroying unwanted material. Higher still, a further development endowed insects with the ability to produce substances that attach to unwanted cells, causing them to aggregate. In a further stage of evolution that saw the arrival of the vertebrates, the system came closer to the complex form we humans have

DOI: 10.1201/9781003272779-17

inherited – and share with reptiles, birds and mammals. It incorporates elements of each of these ancient defence strategies. The result is a set of distinctive bodily responses that protect us against a huge range of threats from viruses, bacteria, parasites and a host of other microbial beasties. Together, these responses are capable of recognising and ridding our bodies of whatever doesn't belong. Where it fails to achieve this, we become ill.

The first line of defence of the system is the physical barrier presented by tissues such as skin and mucus to external substances, which prevent the entry of many unwanted bugs ('pathogens', to be more precise); some, however, have evolved to get through this. Cuts and fissures and the openings of the respiratory tract are well-known entry points. For those that manage to get through, the next line of defence they encounter is the *innate* immune system. As its name implies, this lies ready and waiting at all times. It responds rapidly to a broad range of the more common microorganisms. Molecules on the surface of familiar pathogens are recognised by the system which then mounts an inflammatory response: redness, swelling, pain and temperature rise. These well-known symptoms of infection are caused by the release of various kinds of chemicals which dilate blood vessels, activate pain receptors and attract defensive cells to neutralise or destroy the pathogen.

However, some pathogens, particularly unfamiliar ones, such as the current coronavirus, manage to get past this line of defence. The next level of protection is provided by the *adaptive* part of the immune system. As its name implies, this adapts to whatever arrives – it copes with the unfamiliar. White blood cell (leucocyte) is the generic name given to a family of cells associated with bodily defence. You may have seen some types – neutrophils and eosinophils, for example – listed after your doctor orders a blood test.

Two distinct kinds of white blood cell form a major part of the adaptive immune system. One type, known as B cells (Figure 16.1), produces antibodies (giant protein molecules) targeted at the specific unfamiliar pathogen that has entered the body. B cells move through the bloodstream to wherever the alien particles are and lock on to them, inactivating them. The other type of white cells, known as T cells, has several functions, including killing alien cells and helping B cells produce antibodies.

Remarkably, both B- and T-type cells develop in such a way that they become specific to a particular pathogen, such as the coronavirus, for example. They recognise a particular antigen – such as the spike protein

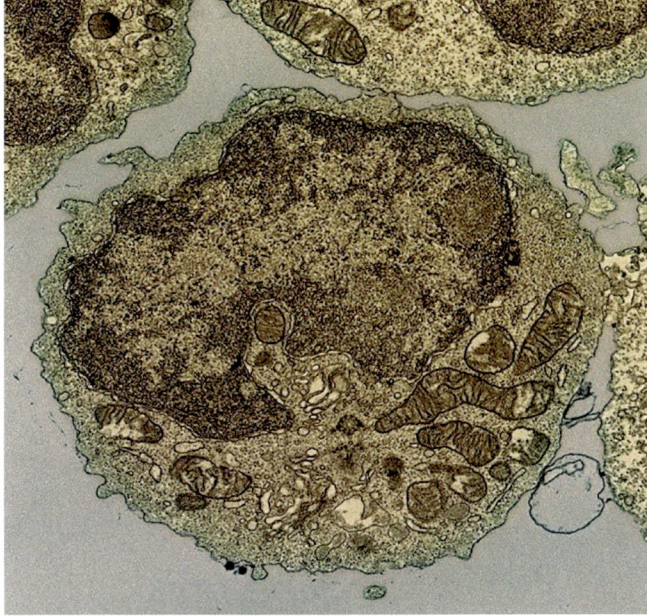

Figure 16.1 Human B cell.
(Image credit: NIAID via Wikimedia, https://creativecommons. org/licenses/by/2.0/deed.en)

on the surface of the coronavirus – and selectively produce antibodies that interact with it. The system adapts itself to cope with the new, unfamiliar threat.

Even more remarkably, a number of these B and T cells help prevent reinfection. They retain the specific antibodies to the particular pathogen, keeping them in reserve, immediately ready to tackle any future occurrence of the pathogen. In effect, the system remembers previous encounters with the pathogen. New-born babies, of course, have no such 'memory'. To protect them, antibodies from the mother are passed through the placenta and milk which last until the baby develops its own during its first few months after birth.

SPECIFICITY

Perhaps the most remarkable feature of the immune system is its ability to recognise almost any horrible bug that's thrown at it: measles, mumps, colds, flu or whatever. Like so many recognition processes in the body, this works thanks to the way protein molecules interact with other smaller structures.

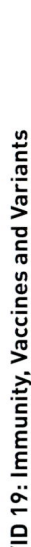

Proteins are large globular molecules with knobbly surfaces pitted with crevices. They have evolved in such a way that particular sites on their surfaces, known as binding sites, fit neatly around particular small molecules with which they interact (Figure 16.2). They are as locks to the keys presented by the smaller molecule (or part thereof), as described earlier in relation to enzymes (Chapter 9). This geometrical 'fit' between the larger protein and the smaller molecule is enhanced by electrical attraction between negatively and positively charged zones around the atoms of each molecule.

This kind of highly specific interaction between large proteins and smaller molecules is not peculiar to the immune system. It underlies the great range of processes that keep our bodies functioning: enzymes snipping up carbohydrates in our intestines, oxygen latching on to haemoglobin in the lungs, as well as antibodies singling out unwanted pathogens, such as coronaviruses.

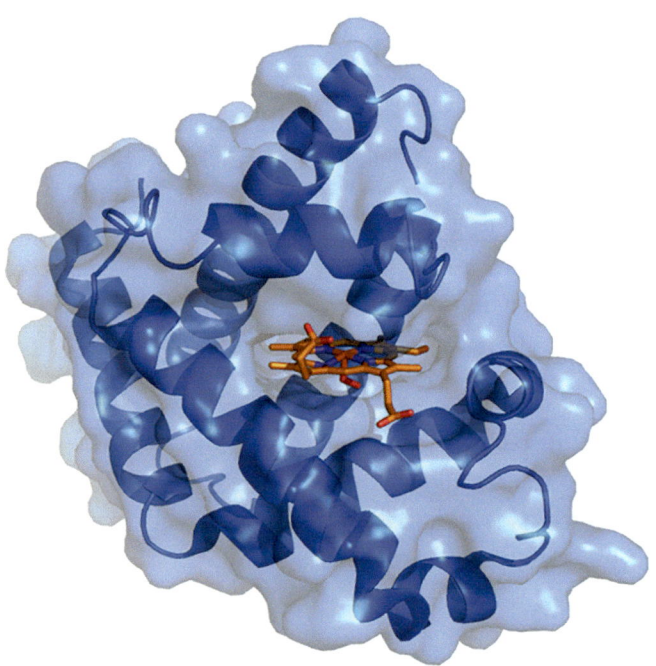

Figure 16.2 Example of a protein molecule (blue) interacting with a small molecule (red).
(Image credit: Thomas Splettstoesser via Wikipedia, https://creativecommons.org/licenses/by-sa/3.0/deed.en)

ANTIBODIES

Antibodies are an extraordinary type of giant protein molecule. Like other types, they have their special sites, their crevices, shaped to fit the smaller molecules to which they must attach. But unlike other proteins they must be capable of 'recognising', and binding to, smaller molecules (or parts of big molecules) of a vast number of different kinds – bits projecting out from all the various unwanted kinds of bacteria, viruses, parasites and fungi, for example.

To achieve this enormous flexibility, a unique architecture has evolved within the family of antibody molecules (known technically as immunoglobulins). In the model in Figure 16.3, each little blob represents an atom. Linked together, the many thousands of atoms constitute the giant Y-shaped antibody molecule. Two chains of atoms (blue and yellow) run from top to bottom. A narrow hinge region in the middle confers enormous flexibility to the molecule around its waist. Two further chunks (green and pink) form part of the upper two arms of the Y-shaped molecule.

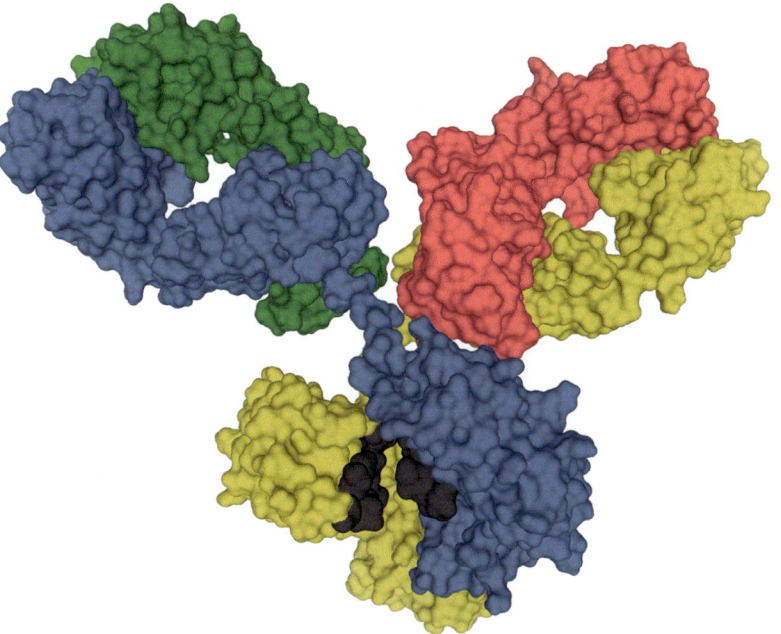

Figure 16.3 Antibody molecule (or immunoglobulin).
(Image courtesy of Tokenzero, https://creativecommons.org/licenses/by-sa/4.0/deed.en)

At the tip of each arm of the Y is a section of the molecule whose precise shape is remarkably variable. Called the binding site it is able to adapt its shape to fit any of the countless pathogens to which it has been exposed. The diagram in Figure 16.4 illustrates the way in which Y-shaped antibody molecules interact with specific antigens. The environment of the antibody's binding sites provides a good fit for just one specific type of antigen (the yellow one in the diagram). Other, differently shaped, antigens (other colours) do not fit.

Each type of virus, bacterium or other pathogen presents a different set of antigens on its surface. The task of the immune system is to ensure that a stock of antibodies with just the right antigen binding site lies ready and waiting for any unwanted antigen that might float past. In the case of coronavirus, the antibody recognises and latches on to a bit of a protein sticking out on the surface of the virus, known as the spike protein. As Figure 16.5 shows, by binding in this way to the spike proteins (red), the antibody molecules (green) prevent the virus attaching to the cell.

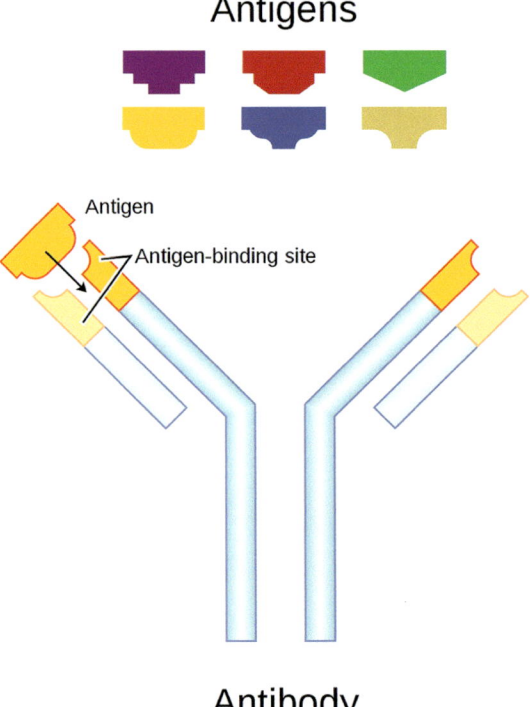

Figure 16.4 Diagram of an antibody and six different antigens.
(Image credit: Fvasconcellos via Wikimedia, Public Domain.)

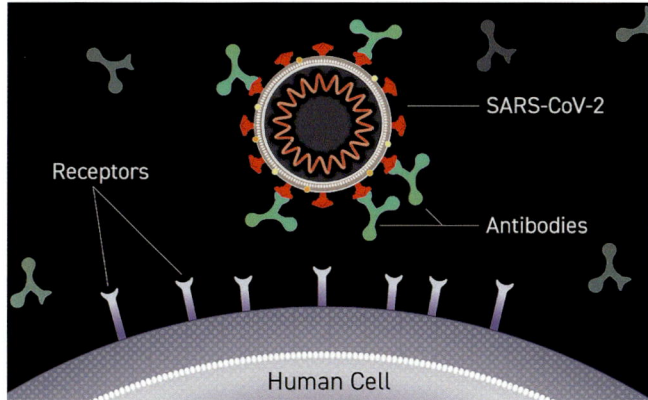

Figure 16.5 Antibodies binding to the surface of a virus, blocking entry into a human cell.
(Image credit: Lisa Donohue, CoVPN, at NIH.)

NATURAL IMMUNITY

It's the extraordinary chemical memory of the immune system that accounts for the well-known experience of immunity after previous exposure to diseases such as mumps or measles. Antibodies and white cells remain stored in the body for long periods of time – up to a lifetime in some cases. If the pathogen reappears at a later date, these specific memory cells of the immune system recognise it and multiply rapidly to eliminate it. It's common experience, however, that it doesn't always work as neatly as this. The degree of immunity can wane over time and, as we know only too well from seasonal influenza, a virus can mutate, becoming unrecognisable to the memory cells. The coronavirus responsible for COVID-19 is indeed mutating and immunity has been shown to wane over time, as reinfection in some patients demonstrates.

VACCINATION

Immunity can also be induced artificially, as well as naturally following a bout of infection – that's what vaccination is all about. The vaccine causes the antigen – a fragment of the spike protein in the case of coronavirus – to appear in the bloodstream. This stimulates the immune system to mount a defence, just as it would do if the infection had occurred naturally. It develops specific B and T cells and antibodies capable of responding to the new pathogen. These multiply as they would for a natural infection and ultimately destroy the material inserted via the vaccine. As described above, copies of the specific B and T cells are retained as a kind of 'archive', ready to mobilise resistance to any future encounter with the virus.

In the case of coronavirus, different vaccines work in different ways. Some well-known ones do not in fact contain the spike protein itself or indeed any other antigen. Instead, they contain the genetic code (RNA or DNA) for making the spike protein and this is injected into the body inside a protective coat. This genetic message finds its way inside our cells where it is used by the normal machinery of the cell to produce large numbers of copies of the spike protein (or a mutated version of it). It's these copies of the viral protein, made inside our bodies, that go on to stimulate our immune systems.

A key challenge for vaccine developers is to find a way to get the RNA or DNA – the genetic information for the spike protein – into our cells. Once it has reached the target cells in the respiratory tract, it has to be able to penetrate the outer membrane of the cell, a key purpose of which is precisely to keep out unfamiliar molecules and particles. A variety of ingenious ways have been devised by vaccine developers to achieve this break-in. The AstraZeneca, Johnson & Johnson and Sputnik vaccines cunningly exploit a special quality of all viruses: the ability to penetrate cells. It's what viruses do! Vaccines of this type make use of a virus that causes the common cold in chimpanzees – nothing to do with coronavirus. They use a weakened version of this harmless chimp virus to penetrate our body's defences and enter our cells. The vital genetic information needed to produce the antigen is loaded inside the helpful chimp virus. Once inside our cells, the chimp virus deposits its genetic cargo but does no harm itself. Our cells then make use of the RNA or DNA to manufacture the coronavirus spike protein. The AstraZeneca version uses a type of RNA, called messenger RNA (mRNA), to convey the genetic information for producing the spike protein; the Johnson & Johnson version uses the corresponding DNA. Both methods enable the spike protein, or a mutated version of it, to be synthesised within the cells, ready to enter the bloodstream.

The Moderna and Pfizer-BioNtech versions, on the other hand, use an artificial nanoparticle – an oily globule – instead of a weakened chimpanzee virus to get inside our cells. The model in Figure 16.6 shows such a nanoparticle enclosing the vital genetic material from a virus – the green spiral molecules. The blue, purple and orange spheres with protruding tails are various oily lipid molecules which protect and support the valuable cargo.

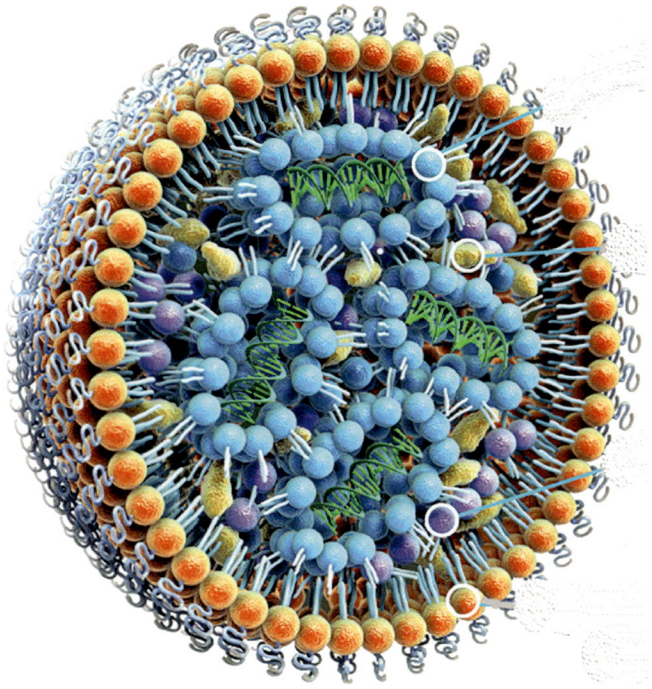

Figure 16.6 Vaccine nanoparticle enclosing the vital genetic material from a virus.
(Image credit: ngodler@precision-nano.com)

A very different type of vaccine does not use RNA or DNA at all. Instead of inserting genetic material into our bodies, the Novavax version uses a more traditional approach in which the spike protein itself is injected. This provokes the immune system directly into developing antibodies. Whichever way the spike protein gets into our bloodstream, it causes the specific antibodies needed to combat the coronavirus to be created. Crucially, copies of them get stored away – hopefully for a long time – ready to pounce, were the infection to arrive at some future date.

MUTATIONS AND VARIANTS

As we've seen above, the proteins of the immune system are giant, globular-shaped molecules. This apparent shape is deceptive because the molecules are, at the primary level, long thin thread-like structures made up of subunits known as amino acids. The chain is held together wherever two particular subunits, known as cysteine (CYS), come together (Figure 16.7),

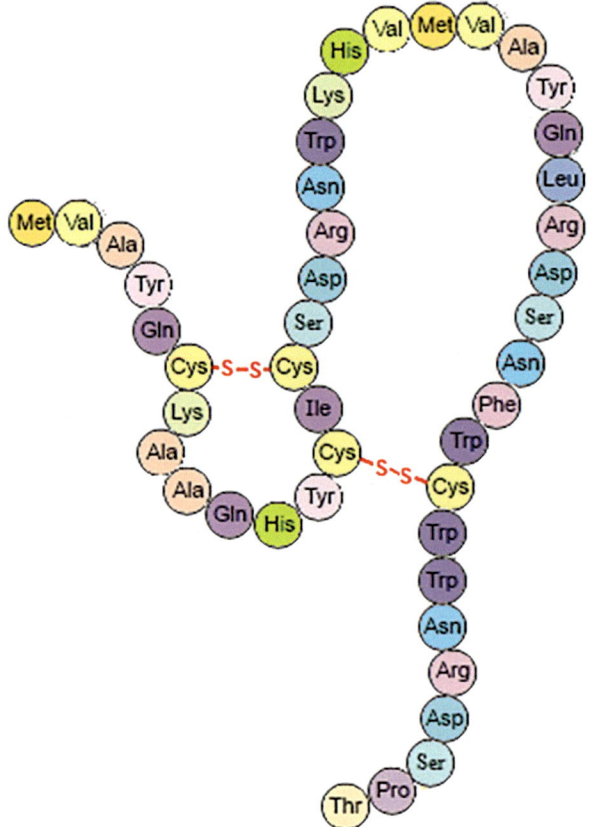

Figure 16.7 Primary structure of a protein – a chain of amino acids. (Image credit: CKRobinson at Wikimedia, https://creativecommons.org/licenses/by-sa/4.0/deed.en)

Due to electrical attraction between some sections of the thread, the long thin molecule wraps up into a globule, rather like a long, beaded necklace held in the palm of the hand. The thread is a sequence of subunits linked together. The precise sequence of these subunits is determined by information coded in the gene for that protein. This is how living systems normally function – proteins are continuously being made using the genes encoded within the molecules of DNA or RNA; every time a cell divides into two, its genetic material has to be duplicated in advance.

But very occasionally mistakes occur as genes are duplicated for transmission from one generation to the next. One of the 'letters' in the genetic code may be misread, or a random collision with a high-energy particle (in cosmic radiation or X rays, for example) may transform one chemical sub-unit ('letter') into another. Such changes are called mutations – from the Latin word for alteration. If the code is changed, the protein which is based upon it may also be changed, as illustrated in Figure 4.4, where we encountered mutations in relation to genetic disorders. Fortunately for us, when such changes – or mutations – do occur, they usually make little or no difference to the behaviour of the protein. Just occasionally, however, they do.

Once a gene has been altered, whether by miscopying or radiation effects, the alteration gets reproduced when the gene replicates. In larger organisms with long lifecycles, such as us humans, this transfer of a mutation from one generation to the next will only occur once every few years or decades. Viruses, on the other hand, bring forth the next generation in a matter of minutes. Many generations are passed through in the course of a day and each occasion introduces the chance of a mutation. That's why it's so important, in an epidemic, to bring down the overall amount of virus in a population as quickly as possible, to minimise the number of replications, each with its attendant risk of a mutation that might turn out to have unwanted consequences for us.

As with evolution generally, different versions of a virus will survive and reproduce with different degrees of success. Those that survive longest and produce the largest number of offspring will eventually outnumber the others in the population. With such rapid turnover, new versions of the virus that are well fitted to their environment will soon spread amongst the general population of viruses. The COVID-19 pandemic has illustrated this evolutionary point clearly, as one mutant version after another has come to dominate – from alpha to beta to delta to omicron, at the time of writing.

Over time, the genetic material of a virus will acquire more than one mutation at various points along its sequence of many thousands of bases (or 'letters'). The word 'variant' is usually used to denote a version of the virus containing several mutations. If a variant becomes established and starts behaving in a distinctly different way, it may come to be called a 'strain' (though the terminology is not precise).

At the time of writing (Autumn 2021), thousands of distinct mutations have been identified – i.e., locations on the viral RNA where a base (i.e., a single letter in the code) has been altered. This is entirely normal and the vast majority cause no harm. The number of variants – each with its characteristic set of mutations – that are viable and able to spread is much smaller. The effects they each have on disease symptoms are closely monitored as and when they are identified. The colourful chart in Figure 16.8, maintained on a daily basis by the international open-source collaboration *Nextstrain* (https://nextstrain.org/ncov/global?m=div), shows each 'clade' (virus family descended from a common ancestor) in a particular colour. Each blob represents a unique genetic sequence identified in a laboratory somewhere in the world. The number of mutations in each case is indicated along the horizontal axis. There are hundreds of thousands of these slightly differing sequences. Because the geographical location at which each mutation was first identified is recorded and shared globally, the journey made by each specific variant as it crosses the world can be plotted. This is how we're able to trace patterns of transmission across the world.

The chart shows the situation in October 2021 and gives a sense of how variants are tracked. An up-to-date picture is available on the Nextstrain website: https://nextstrain.org/ncov/gisaid/global?c=num_date&m=div

The ability of vaccines to respond to each of the variants is a matter of continuous research in real time around the world. Fortunately, the latest method of delivering vaccines, by enclosing the specific gene for the spike protein in a generic particle capable of entering our cells, helps cope with these variants. It means that the gene for a new variant can be inserted into the original, already-tested delivery particle, thus speeding up the time taken to update the vaccine.

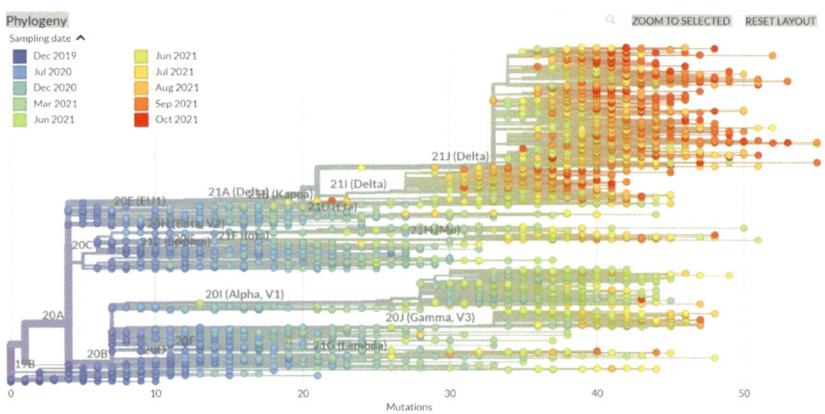

Figure 16.8 Genetic sequences of coronaviruses, organised by family.
(Image credit: nextstrain.org, CC-BY-4.0 licence)

CONCLUSION

This chapter and the last have covered a lot of ground and touched on many different areas of science, from virology and anatomy to epidemiology and behavioural science. The pandemic has brought scientists to our screens at peak time, more than ever before. Behind the visible experts lie the legions of engineers working on ventilation dynamics, physicists designing vaccine delivery particles, mathematicians calculating risk and social scientists analysing the response of specific groups. An extraordinary consequence of this worldwide affliction has been to bring the general public much closer to the reality of science as it actually works in real time. Uncertainty and contradiction have been made plain, and hopes and fears have been shared. Important features of a healthy scientific culture have been clearly demonstrated. Data are gathered carefully and analysed sceptically; individuals work together in multi-skill teams, specialists collaborate across disciplines, information is shared across the globe. Crucially, knowledge accumulated through decades of basic research is applied to the questions of the moment.

Also revealed by the exigencies of the pandemic has been the interaction of science with political, economic and social affairs. Far from the certainty and rigidity so often (and inappropriately) associated with scientific knowledge, the unfolding of the pandemic has highlighted the provisional and contestable nature of research findings. To be useful to the public at large, multiple studies, carried out in varying contexts, using differing methods for differing purposes, have to be aggregated and distilled. Even then, different interpretations can be legitimately put upon the same results, especially when they are partial or have large margins of error.

Preparing and communicating clear messages from a mass of complex, and sometimes conflicting, scientific information is no simple matter. Yet doing so is essential at a time of crisis and will continue to be so as we respond to ongoing environmental, as well as public health, threats. It is encouraging to witness what appears to be growing interest in science communication, both on the part of the scientific community and amongst the public in general.

Energy and the Climate Emergency
Greenhouse Gases, Energy and Heat Pumps
Seventeen

The future of our planet is the overwhelming concern of our day. Extreme threats are piling up from multiple directions: warming of the atmosphere, diminishing diversity of species, a rising tide of plastic waste and falling supplies of fresh water. Considered as a whole, the prospects are daunting, to put it mildly. For many, the response is to turn away, feeling intimidated by the scale of the problems and powerless as an individual to respond to them. It is impossible to foresee how the world's populations and leaders are going to respond over the coming decades, but some things are clear even now: we need to try, as individuals, to understand what is happening and, as communities, to work together to minimise the degree of warming and mitigate the foreseeable consequences.

In this chapter, we look at just one of these interconnected issues, global warming, and focus on an aspect that affects us all directly: energy. Science has of course been at the very forefront of the global warming story. As long ago as the late nineteenth century, scientists saw that the level of so-called 'greenhouse gases' in the atmosphere could affect cooling and warming at the surface of the Earth. Studies of the ways in which gases from both natural and human sources interact with the oceans and the atmosphere continued during the twentieth century culminating in a key 1965 report *Restoring the Quality of Our Environment* by the U.S. Science Advisory Committee. It warned of the 'harmful effects of fossil fuel emissions. An increase of atmospheric carbon dioxide could act, much like the glass in a greenhouse, to raise the temperature of the lower air' and declared 'the rise of atmospheric carbon dioxide levels to be the direct result of fossil fuel burning'.

Research across multiple disciplines intensified during the 1970s and 1980s, and findings were brought together at the founding of the UN's Intergovernmental Panel on Climate Change in 1988. Its first report in 1990 marked the development of a scientific consensus, expressed clearly in its 1995 report: 'The balance of evidence suggests a discernible human influence on global climate'. It has taken a long time for the grim

DOI: 10.1201/9781003272779-18

reality implied by these IPCC reports to break through to the consciousness of political and business leaders and to populations at large. Now the race is on to secure reductions in the emission rates of greenhouse gases such as carbon dioxide and methane.

GREENHOUSE EFFECT

It's not at all easy to visualise why using your air-conditioning unit or eating beef helps to warm up the planet. Perhaps this problem of imagination is one reason the world has taken so long to wake up to the danger.

The key scientific concept is the 'greenhouse effect'. Inside a greenhouse, it's warmer than outside because the heat energy passing into it through the glass comes from the very high temperature Sun, whereas the heat energy passing back out of it through the glass comes from the much cooler flooring and objects in the greenhouse. Heat radiates from objects in the form of waves of various frequencies. The hotter the object, the higher the frequency of the waves it radiates. Because the Sun is so hot, a high proportion of the waves radiating from it have a relatively high frequency. Such waves are able to pass easily through a substance like glass; but lower frequency waves are absorbed to a greater extent as they pass through it. Radiation from the relatively cool contents of a greenhouse is of a lower frequency and, as a consequence, escapes the greenhouse less effectively. So heat energy enters the greenhouse more readily than it leaves, causing the interior to warm up.

Although glass does not surround our planet, some gases have a similar effect to glass. As illustrated in Figure 17.1, they allow high frequency radiation from the Sun to pass through to the surface of the Earth (yellow lines in the diagram), but absorb more of the lower frequency waves emanating back outwards from the surface of the Earth (red arrows). Once these so-called greenhouse gases have absorbed this energy from the Earth, they re-radiate it out in all directions. Some of this is directed straight back to Earth, as shown by the red lines in the diagram. In this way, a proportion of the heat energy originating from the Sun bounces back and forth in the Earth's atmosphere, effectively trapped. It's as though a blanket encloses the Earth.

The diagram also indicates that gases differ in the degree to which they absorb the heat radiating up from the Earth. Carbon dioxide (CO_2), methane (CH_4) and water vapour (H_2O), for example, absorb significantly and are called 'greenhouse gases'; oxygen (O_2), nitrogen (N_2) and argon (Ar) do not. Although water vapour is a major greenhouse gas, its

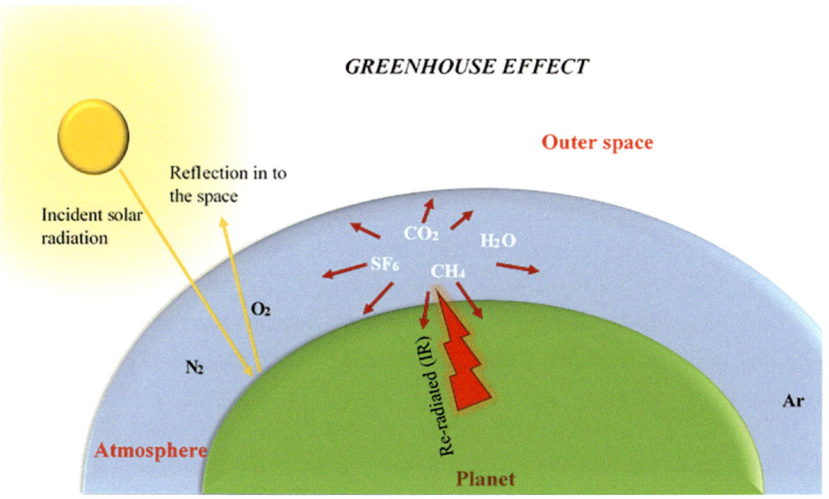

GREENHOUSE EFFECT

Figure 17.1 The greenhouse effect.
(Image courtesy of Wikilakz, https://creativecommons.org/licenses/by-sa/4.0/deed.en)

concentration in the atmosphere is not greatly affected by human activity. Methane is also a major greenhouse gas, but currently is present in relatively low concentrations (though it's much more potent than CO_2). The main greenhouse gas in the atmosphere contributed by human activity is carbon dioxide. This arises mainly from burning coal, oil and gas, known collectively as fossil fuels. These were estimated to constitute 62% of greenhouse gas emissions in 2015. At the time of writing (2021), coal-fired power stations contribute a fifth of the total emissions globally.

SOURCES OF GREENHOUSE GASES

Fossil fuels, as their name implies, were formed over millions of years from the remains of ancient living things: plants and animals. Plants had captured energy from the Sun's rays by the process of photosynthesis. This solar energy had enabled carbon dioxide in the air to combine with water via the roots in a chemical reaction. This produced the high-energy, carbon-rich molecules of which plants are made, including sugars, starch and cellulose. This is what is really meant by the phrase 'capturing energy': a continuous stream of energy radiating from the Sun – in a form known as electromagnetic energy – ends up being stored over time in a different form – chemical energy – in the molecules that make up the plant. As a result, burning of fossil fuels provides not only something we really want – energy – but also something we seriously don't want: carbon dioxide. As an aside, the process of photosynthesis also produces oxygen which is released into the air, enabling creatures, including ourselves, to live and breathe.

Burning (or combustion) is a process that reverses the work of photosynthesis. It breaks up the energy-rich carbohydrate molecules that make up plants, consuming oxygen from the air as part of the chemical reaction. On the one hand, this process releases stored-up energy for us to use but, on the other hand, it creates the greenhouse gas carbon dioxide by causing carbon atoms in the carbohydrate molecules to combine with oxygen from the air.

It's not only combustion that is causing the level of unwanted gases in the atmosphere to rise. The decreasing size of forests across the globe is another important cause. Living plants take in carbon dioxide from the atmosphere today, just as their ancestors did long ago in creating the fossil fuels. As trees are cut down, the quantity of CO_2 extracted from the atmosphere by the diminishing forests decreases; the atmospheric level rises.

Agricultural methods also contribute to greenhouse gas emissions – somewhere between 10% and 17% of the global total according to the IPCC. When fertilisers containing nitrogen are spread on fields, they lead to the emission of another powerful greenhouse gas: nitrous oxide. The fertiliser encourages the growth of bacteria that produce the gas and also reacts directly with chemicals in the soil to produce the gas. A third serious greenhouse gas – methane – is the unfortunate product of the way cows digest their food. Microbes in the guts of ruminant animals, like cows, sheep and goats, break down molecules like cellulose in grass and other plants that we humans are unable to digest. This process is a kind of fermentation which, in the absence of oxygen, produces methane gas which animals belch out or otherwise expel.

Production of cement is another important process in which greenhouse gases are emitted on a grand scale, accounting for 7%–10% of global carbon dioxide emissions. In this case, however, it is mainly a chemical reaction, inherent in the process of manufacturing cement that is responsible. Cement is made by heating limestone, a kind of rock found across the world, made of calcium carbonate. As this name implies, this contains carbon and this gets removed during the heating process in the form of carbon dioxide. Total emissions by the cement industry include not only this but also emissions from burning the fuel used to heat the limestone.

As we have seen, several kinds of gas are contributing to the greenhouse effect that is warming the planet and these are emitted in the course of a huge variety of human activities. We have highlighted just a few of the major kinds.

WHAT IS ENERGY?

We have seen that the greenhouse effect is, in effect, upsetting a vital energy balance that makes life on Earth possible. The flow of energy impinging on our planet from the Sun has to be balanced by an equal flow of energy away from our planet if the temperature here on the surface is to remain steady. This is of course the case for any steady situation: it's how our bodies maintain a steady temperature and, if we're lucky, how a thermostat does the same for our rooms. But currently, and over recent centuries, the outflow of energy has been gradually reducing, thanks to the blanket encircling the globe.

It's at this point in a discussion one evening that a participant, Helen, posed a fundamental question, somewhat hesitantly: 'What actually is energy?' We had considered energy radiating from the Sun, being stored in plants and consumed as food or fuel. She was aware of how frequently she comes across the word: in her electricity bills, on food labels and in political debates, yet had no idea what it actually was. 'Is there a difference between energy and force?' she asked tentatively.

Etymologically, the word 'energy' (from the ancient Greek) is associated with activity or work, whereas 'force' is linked to strength (from Latin). The scientific distinction is quite close to the everyday difference you might perceive between a person who is energetic and another who is forceful. Force refers to a push or pull on an object that can speed it up or slow it down, or make it change direction or change its shape. It's fairly easy to picture so long as things are in contact – like pulling on a rope, or hammering in a nail – but when a force acts on things remote from one another, it calls for greater imagination. When you first see a magnet attracting a pin some distance away, as a child, or notice your comb clinging on to fine dry hair, it all seems magical. Science has developed the idea of a 'field' to capture the zone in which a force exerts its influence. This is what is implied by the phrase 'magnetic field' or 'gravitational field'.

Our concept of energy is quite a different matter. Energy is an abstraction used to keep an account of changes we observe in systems. If that sounds vague, it's meant to. Energy is notoriously difficult to define as a generalisation. The physicist Richard Feynmann is quoted as saying 'in physics today we have no knowledge what energy is'. It's more readily described in particular situations – that's how the concept is usually taught at school.

There are the *forms* of energy: heat energy, nuclear energy, chemical energy, electrical energy, kinetic energy, potential energy… but what is it that makes them all 'energy'? The formal definition in physics is 'the capacity to do work' which means the capacity to make a force move through some distance. This is easy to imagine in a mechanical context like the force driving a piston up and down in a car engine or the work you do in pushing a child's swing through a certain distance. Energy in these contexts refers to the capacity, locked up in a fuel, to drive a piston or, in a parent (sometimes), to push a child on a swing.

It's less easy to conceptualise in the chemical, electrical or nuclear case. In these situations, we are aware of energy as it manifests itself in a transformation. A jet of gas bursts into flame on your hob, a vacuum cleaner jumps into action or a bomb explodes. This idea of transformation underlies the deeper meaning of energy: it is an abstract quantity – a number – which remains unaltered throughout any transformation. This is captured in the well-known law of the conservation of energy, which you may recall being drummed into you at school: 'energy is neither created nor destroyed'. Trying to conjure up a physical image for energy is fruitless: it's a mathematical abstraction. Like the concept of money, it may take on physical appearances like cash or numbers on a bank statement, but in the end, it's an abstract human invention. Unlike money, however, energy is always conserved!

ENERGY TRANSFORMATIONS

To get an intuitive feel for energy, we need to look at particular manifestations of this abstraction. As touched on above, a fundamental energy transformation occurs within plants when light energy radiating from the surface of the Sun falls on leaves and propels the process of photosynthesis. In this process, electromagnetic energy in the form of light gets locked up as chemical energy in carbohydrate molecules created when water and carbon dioxide react together in the cells of the plant. Prior to this, the electromagnetic (light) energy radiated from the Sun had to arise from an even earlier transformation. In this case, it was nuclear reactions occurring in the extremely hot and pressurised core of the Sun that transformed nuclear energy to electromagnetic energy. When this reached the Sun's surface, it was radiated away, as light and heat (and other kinds of radiation), some of it in the direction of Earth.

Energy locked up in plant material today provides the fuel that sustains our lives and those of other creatures. By digesting energy-rich carbohydrates from fruit, vegetables and staple crops, we transform the chemical energy locked up in them into mechanical energy through our muscles, enabling us to move and lift, and into the heat energy needed to maintain our elevated body temperatures. We may also capture that energy by consuming meat from animals that have themselves fed on plant material.

Energy transformed by humans and animals into its mechanical form was the basis of economic life in the pre-industrial period. We – or our horses, mules, camels or elephants – laboured in fields, struggled up hills or ground up our cereal grains using muscle power alone. The Industrial Age saw the concept of energy develop through the efforts of innovators in manufacturing and transportation and by an emerging community of scientists. Ways were found to transform energy locked up in fossil fuels into heat energy and thence, through the production of steam, into mechanical energy to drive machines. Pistons were forced by high-pressure steam entering the cylinders of the engines that turned weaving machines, pumped out coalmines and spun the wheels of railway locomotives. Understanding these transformations and how to make them efficient not only drove rapid industrial innovation throughout the nineteenth century but also opened up a new theoretical subject linking the previously disconnected worlds of mechanics and heat: thermodynamics.

Today, however, the long-term cost of this era of intense industrialisation is plain to see. The carbon dioxide gas produced as a by-product of burning fossil fuels has gradually warmed our planet and led to the catastrophic consequences for our climate of which we are aware today. In the coming decades, energy is going to have to come from other sources, however hard this will be to achieve. Different kinds of transformation will be required.

ALTERNATIVE SOURCES OF ENERGY

Paradoxically, abundant quantities of energy, in many forms, surround us in the natural environment. Waves at sea rise and fall continuously, tossing boats up and down. Tides drive great masses of water up estuaries in a twice daily rhythm. Winds with energy sufficient to uproot trees come and go and sunshine warms our rooves, fields and deserts. Energy from these alternative sources will need to be captured, transformed and stored for use as fossil fuels are phased out.

Devices to make this happen are continuously being developed and policies to promote them gradually emerging. The machines produced by the transformative technologies look very different from one another – wind turbines, solar panels, tidal barrages, wave generators. What they have in common is that they do not involve burning, and the consequent emission of greenhouse gases. They capture energy in mechanical form, like the movement of air in the wind or water in waves, or in the form of heat, as with heat pumps or solar panels, and transform it into its electrical form. Thereafter, electricity becomes the form in which energy reaches our houses, drives our vehicles and powers our machinery.

The means by which these transformations are made to happen are as different from one another as waves are from sunshine. To illustrate how such devices work, we conclude this chapter by focussing on one or two changes we can expect to see in our houses over the coming years.

HEAT PUMPS

How times have changed! Heat pumps were once mentioned only in the remote recesses of a degree in physics. Today, they are common parlance amongst plumbers, home improvers and increasingly our political representatives – a good thing too (Figure 17.2).

What baffled people in a science group discussion was the seemingly nonsensical idea that, as Wendy put it: 'you could get heat from the cold earth or cool air to heat your home'. Yes – that's exactly what heat pumps do; indeed, it's why they are called 'pumps'. A pump for your tyres will take air from a low pressure environment (the atmosphere) and force it into a high-pressure one (inside a tyre); a Dutch windmill will take water from low-lying fields and pump it, against gravity, up to sea level. By analogy, a heat pumps take heat energy from a low temperature environment and 'pumps' it up into a higher temperature one.

At the heart of the conceptual difficulty lies an important issue in physics: distinguishing heat from temperature. This is explored in Chapter 12 'A Nice Warm Shower'. The key point is that 'heat' is a form of energy associated with the molecules or atoms of which a substance is composed. It's the mechanical energy associated with the motion of molecules. All molecules are moving to some degree; they may be rushing around or spinning or vibrating. The faster they spin or vibrate or rush around, the greater their energy. This understanding, developed in the nineteenth century, enables us to see that there is a certain amount of heat energy in a block of ice or a zephyr of wind just as there is in a

Figure 17.2 A heat pump.
(Image credit: Ppntori via Wikimedia, Public Domain.)

tub of molten steel. Of more significance today, there is heat energy in the earth and the air around us, however warm or cool they may be.

The temperature at which heat energy is available varies of course. Temperature is a measure of the *average* level of energy in the zillions of molecules of a substance. In a warm room, the molecules of air are moving around faster, on average then in a cooler one. The task for a heat pump is to take in a quantity of heat energy from one place and expel heat energy at a higher temperature in another place. To do this, molecules have to be energised – speeded up.

A heat pump achieves this by using the same principle as, and similar technology to, a refrigerator. The latter extracts heat energy from within its cabinet and pumps it out at a higher temperature through the black tubing at the back. It does this by employing a mediating substance that cycles around, day in, day out, through a circuit of tubing inside its walls. The mediating substance (known as a refrigerant) is a volatile liquid, i.e., one that can change easily from liquid to vapour form, just as bottled fragrances do.

A heat pump carries out a similar operation as a fridge. It also extracts heat from a cooler place (the outdoors) and dissipates it in a warmer place (indoors) at a higher temperature.

The task of taking heat energy from a cooler place, raising its temperature and pushing it out in a warmer place, is made possible by the use of the refrigerant – a fluid that circulates through a circuit of narrow tubing. The key processes involved are evaporation and condensation of the refrigerant. We know from everyday experience that when a volatile liquid like eau-de-cologne evaporates off our skin, it cools us. Evaporation extracts heat energy from its surroundings (our skin in this case). Conversely, when a vapour condenses into the liquid state, as when a cold window mists up, heat energy is released into the surroundings. The refrigerant circulates around and around picking up energy in one place and dissipating it in another – it's a kind of transporter, like an endless airport conveyor belt, taking on luggage in one place and offloading it elsewhere.

A more detailed and explanation follows which can be hard to grasp first time. It can be skipped without loss, on first reading.

In a heat pump, the refrigerant is sometimes in the liquid state, sometimes in the vapour state as it circulates – and sometimes transitioning between the two.

Starting at the bottom of the diagram in Figure 17.3, the refrigerant, as a vapour, enters a compressor (4) where it gets compressed, much as the air does in a bike or car pump. This act of compression, driven by electrical power, raises the temperature of the vapour – just as it does when you pump up a tyre. The hot vapour leaving the compressor circulates round to a place where it gives up much of its heat energy indoors (the condenser, 1) by passing though tubing exposed to the air in the room to be heated (or to water pipes that carry the heat away to where it is required). As this heat energy is gradually dissipated, the hot vapour cools and condenses into its liquid state, though it remains pressurised.

At the next point in its journey, the refrigerant (now a liquid) passes through a narrow opening or valve (2), which allows the liquid to decompress as it moves into a wider space. This act of decompression enables the liquid to evaporate partially – like a deodorant passing through the nozzle of a can. This lowers the temperature – the reverse of the compression stage which raised the temperature. The resulting cool mixture of liquid and vapour circulates on through the pipes to the outside environment, where it has by now become cooler than the

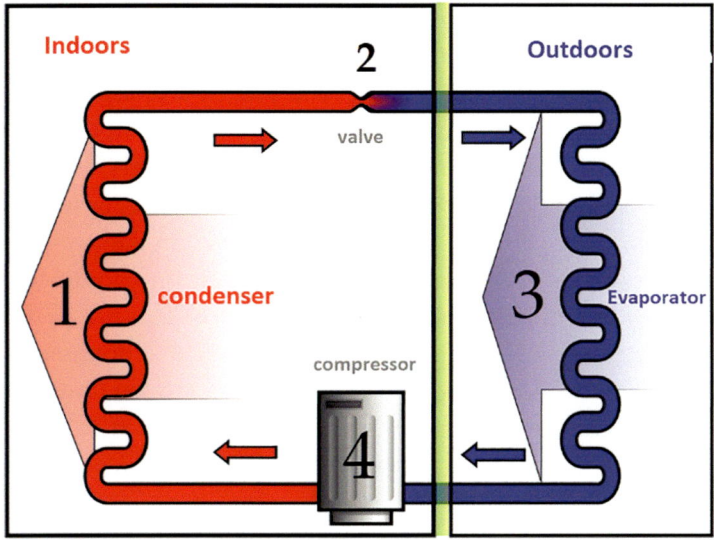

Figure 17.3 Operating cycle of a heat pump.
(Image credit: Ilmari Karonen via Wikimedia, Public Domain.)

surroundings – the ground or the air. As a consequence, heat energy from the slightly warmer earth or air enters into the piping, warming up the cool refrigerant mixture. The warmed-up refrigerant circulates on, back into the compressor to repeat the cycle.

In summary, heat energy extracted from outdoors has its temperature raised by an act of compression. This higher temperature heat energy has been pushed out indoors to warm up an interior. The mediating substance – the refrigerant – is then cooled by an act of decompression in readiness to pick up more heat energy from the exterior. A heat pump does consume some electrical energy to run the compressor, but it's less than would be used to heat a house by purely electrical means.

ENERGY LOSS

Raising the temperature of a room in cool weather, as we all know, doesn't come cheap. Nor does lowering it, in hot weather. Currently, we pay for it through the bills we receive for fuel to run our boilers and air-conditioning units. As fossil fuels are gradually replaced by alternative sources of energy, we will still have to pay for energy. Heat pumps and air-conditioning units consume electrical energy to drive their compressors and this has to be produced somehow. If we are to reduce our overall consumption of energy in the years to come, we also need to deal with the way heat energy is lost so rapidly from our buildings. Improved insulation will not only reduce our bills, it will also enable comfortable room temperatures to be achieved with heat pumps.

Heat losses occur because whenever the temperature differs between two places; heat energy flows from the higher to the lower, never the reverse. As the comic duo, Flanders and Swan expressed it in their version of the second law of thermodynamics 'heat won't pass from a cooler to a hotter; you can try it if you like but you far better notter'. This is a matter of everyday experience – a hot drink always cools down in a room, never heats up spontaneously. To maintain a room at a steady temperature, the rate at which heat energy enters it has to equal that at which it leaves. By reducing the rate of heat loss, the rate at which energy is consumed is also reduced.

Building design in the era of climate change will require losses to be minimised. Old buildings will need to adapted and new ones thoughtfully designed. Fortunately, the science of heat transfer is well understood and not too complicated. As described in Chapter 12 in connection with baths, it is transferred in three distinct ways: conduction, convection and radiation. It's by impeding the loss of heat through each of these means that we can reduce our energy bills and contribute to healing the planet.

Heat energy is conducted through materials at a rate which depends on both the nature of the material it passes through and the difference in temperature on each side or end. The energy is physically passed from one molecule or atom to the next in a substance. As we saw in Chapter 12, each of these is in continual motion – vibrating, rotating or rushing about. The more energetic movers and shakers simply pass on some of their energy of movement to their immediate neighbours by rubbing against them or colliding with them, in a kind of pass-the-parcel exercise. It's how heat travels up the shaft of a spoon left in a cup of tea.

Brick conducts heat fairly well but air much less well. This fact led to buildings being designed, from the late nineteenth century onwards, with two solid walls separated by an air space: so-called cavity walls. Today, the partial barrier to heat conduction offered by the air in a cavity is often strengthened by including solid foam in the air space. As a poor conductor of heat, this material helps reduce heat loss. It adds to this by also reducing losses by convection: the physical movement of whole volumes of gas or liquid from warmer, less dense zones to cooler ones.

Figure 17.4 Foam insulation in a cavity wall.
(Image credit: The Greenage.)

This results in warm air, rising up, to be replaced where it was with denser, colder air. As this then warms up, it too will rise, move into cooler zones, become more dense again and start to fall beside the rising fluid. In this way, a circulating current is set up transporting heat energy form higher to lower temperature regions. Filling a cavity wall with an obstacle like foam or mineral wool impedes this process, reducing the transfer of energy from the interior of a building to the exterior (Figure 17.4).

You may also have noticed that panels of insulating foam inserted into a cavity wall sometimes have a silvery surface (Figure 17.4), just like the silvery sheets placed on car windscreen in hot climates or wrapped around victims of hypothermia. This is aimed at reducing heat transfer by the third means: radiation. This mechanism differs crucially from the other two in that it is not reliant on any material medium to convey heat energy. Indeed, our ultimate source of energy, the Sun, transfers all its heat energy (and light, too) to us through empty space by this means. When exploring the nature of light in Chapter 3, we saw that objects radiate some heat energy in the form of electromagnetic waves (known as infrared) at any temperature – even the ordinary temperature of a room.

Electromagnetic waves travel in straight lines normally, but when they hit a reflective surface, such as a mirror or aluminium foil, they are deflected: a fraction of their energy bounces back. This enables heat losses to be reduced by placing reflective material in the path of the radiating heat. Placed in the walls of buildings or around the body of a hypothermia victim, this helps reduce heat transfer from inside to outside.

CONCLUSION

The opportunities for curbing the emission of greenhouse gases and mitigating the effects of climate change are endless. Innovative technologies for replacing fossil fuels and traditional methods for reducing heat losses abound. Without doubt, ingenious new ideas and techniques will emerge in the decades to come. However, it is already becoming clear that developments in science and technology alone are unlikely to determine the future of our warming planet. The political and economic challenges of financing the changes needed, and the willingness of people to adjust to them, are enormous. How policies develop on a worldwide basis is quite unforeseeable. New ideas and conflicting interests will no doubt feature prominently in the media, in relation to climate action. The aim of this chapter has been simply to help you, the reader, navigate the choppy waters ahead by introducing some of the relevant principles of physics that may have passed you by earlier.

Eighteen

Electrification is the great hope for tackling the rise in global temperatures. Where today we burn a fossil fuel – coal, gas or oil – to drive our machines and motor vehicles, tomorrow the energy needed will have come from a renewable energy source, via a battery or cable. Major geopolitical issues follow from this. Where will the minerals required for batteries come from? How will electrical energy be distributed? What will happen to the fossil fuel workforce? How will trade be affected as oil-fired planes and ships disappear? The social effects of rapid changes in our way of life are equally uncertain. Will people be prepared to change their dietary habits, travel less or pay to insulate their homes?

None of these questions will be addressed directly in this chapter, important though they are – this is not a political or social account. Instead we address basic questions about the science: What is electricity? Where does it come from? How does it reach us? Fundamental ideas about the nature of electricity will be introduced in the hope that you may feel better equipped to follow the political and social changes to come, as we cope with the climate emergency. As this chapter introduces many interlinked concepts, it is inevitably longer than others and more dense. It can be read in stages.

WHAT IS ELECTRICITY?

Such a basic question seems odd for something we are so familiar with, yet it's just where people in discussion groups wanted to start in trying to understand future energy policy and their ever-rising energy bills. Put simply, 'electricity' is a loosely defined word referring to the flow of electric charge. But definitions of this kind don't take us far forward – they simply put the question back a step to what is meant by 'electric charge'? Mary, in one memorable discussion, said bluntly: 'Andrew, every time you mention "charge" you lose me. What is it? My husband talks confidently about charging up the battery in the car, but when I ask him what it actually means he is as clueless as me'.

DOI: 10.1201/9781003272779-19

Charge is a term introduced in the eighteenth century to explain the unexpected behaviour of certain substances when rubbed. Originally, the mineral amber (known as 'elektron' in ancient Greek) was known to attract small particles to itself after being rubbed; today, many plastics do the same. A comb, for example, can attract small pieces of paper to itself after being run through hair. Early experimenters in the eighteenth century described the rubbing process as 'charging' a piece of amber with something they called electricity, by analogy with charging a cannon with explosives, or glasses with wine before a toast.

The true explanation of this effect turned out to be rather surprising: far from adding some kind of 'electric fluid' into amber when it is rubbed, it was discovered later that electric charge already exists within substances. It is inherent in all forms of matter: amber, plastic or anything else. The very atoms of which all matter is composed are themselves made of particles with electrical properties, as we saw in connection with adhesives in Chapter 11. Today, we understand that atoms comprise a mixture of positively charged particles (protons) and negatively charged ones (electrons) in precisely equal numbers. The substances we interact with everyday are generally uncharged or neutral overall because of this perfect balance of equal and opposite charges. Yet, still Mary's questions remain: What do we actually mean by 'charge'?

The clearest manifestation of electric charge, as seen by us today or by the early experimenters, is a force that attracts things to one another, as when a plastic comb pulls bits of paper to itself. Objects that have been 'electrified' or charged exert a force on one another. In some cases, this force attracts objects to one another; in other cases, it pushes them apart. To account for this twofold behaviour, a twofold model of electrical force was devised. The electrical nature of matter was divided into two kinds, such that when two of the same kind were close they would be pushed apart; when unlike kinds were close they would attract one another (Figure 18.1). The names 'positive' and 'negative' were assigned to these two types.

The strength of the force between objects that have been charged (like a comb passed through hair) depended on the degree of rubbing – the more vigorous the process, the stronger the force of attraction or repulsion. The word 'charge' was adapted, not only as a verb, to describe the process of charging, but also as a noun, to quantify the effect. The more objects like combs were rubbed, the greater the force they exerted, hence the greater the charge associated with them. So, in summary, charge is

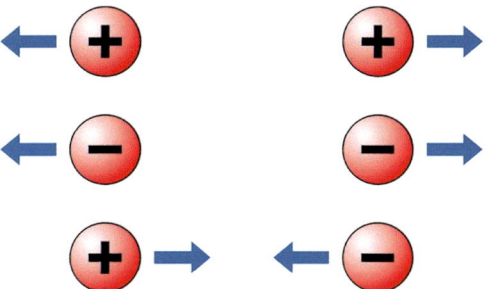

Figure 18.1 Like charges repel; unlike charges attract.
(Image credit: Neuro via Wikimedia, Public Domain.)

Figure 18.2 Electrical attraction caused by rubbing a balloon.
(Image credit: MikeRun via Wikimedia, https://
creativecommons.org/licenses/by-sa/4.0/deed.en)

an abstract quality ascribed to objects to account for an electrically based force that can exist between them.

Static electricity, of the kind described so far, was originally treated as an interesting diversion, a source of entertainment – as it still can be when balloons are rubbed (Figure 18.2). It's only when charges were made to move that electricity became fundamental to our way of life.

MOVING CHARGES

Static electricity, as its name implies, involves electric charge sitting on an object. You can't see it, but may become aware of it as you handle clingfilm or hear a crackle between synthetic fibres. What is of more central importance in our lives is the continuous movement of electric charges through our various machines and devices – what we call a current, by analogy with the flow of water in rivers and oceans. But, as people in discussion groups were eager to ask: What is it that causes charges to flow? How fast do they flow? What stops them? What is it that does the flowing?

The kind of electric current that springs most immediately to mind is that which flows through the wires in our homes, energising our various machines and devices. An equally important kind, however, is the flow of electrical impulses through nerve cells and across the synapses that connect them. Dramatic flashes of lightning are another obvious manifestation and the lights that illuminate our streets and buildings are yet another. Electric charges are clearly able to pass through various kinds of medium: metal in wires, gases in street lamps and, to our peril, through the water in our bathrooms and kitchens. In each of these cases, the charge is associated with a quite different kind of carrier. In gases and water, it's electrically charged atoms (known as ions) that carry the charge; in metals, it's electrons released from atoms that do so. In nerve tissue, it's the rapid movement of charged ions of sodium and potassium in and out of a nerve cell, through its membrane, that enables a signal to pass along. In each case, it's charged particles of one kind or another that do the flowing: ions (charged atoms) in some cases, electrons in others; and other kinds are possible too. What is it that causes these particles to flow rather than just stay put, as they do in static situations?

CURRENT

As we've seen, a charged particle will be attracted to an oppositely charged particle anywhere in the space around it. It's as though a charged particle sets up a zone around itself in which any other charged particle will experience a force of attraction or repulsion. This kind of zone is called a field – an electric field, in this case. At the end of the eighteenth century, Alessandro Volta discovered, more or less by accident, that when a salty solution is placed between two different metals, negatively charged particles (electrons) pile up on one metal, leaving an excess of positive charge on the other. Today, we know this kind of set-up as a battery. Around the charged-up ends (terminals) of a battery, electric fields are created. This can be exploited to make charges flow.

Charges can flow through many kinds of substance – air, water, acids, metals, for example. In other kinds – like rubber or ceramics – they can't. The former are known as conductors, the latter as insulators. Here, we will focus on the flow of charges in metals, such as the wires that bring electrical energy to our homes and workplaces. The basic principles apply equally to movement in air, water and other fluids.

Within metals, including the copper we typically find in wires, atoms are arranged in orderly arrays, like eggs in an egg box (but in three dimensions). In this close-packed arrangement, one or more loosely attached electrons are released from each atom into the surrounding matrix. These delocalised electrons are free to move around. By leaving their previously neutral atoms, the latter become positively charged by an equivalent amount, as indicated in Figure 18.3. Atoms that have lost or gained electrons are called ions.

When a piece of metal, such as a wire, is connected to the negative end of a battery, it is now linked to the excess of electrons sitting on that terminal. As a result, the electric field created by these electrons now spreads throughout the length and breadth of the attached wire. This electric field exerts a force on the delocalised electrons in the wire, tending to drive them away. Equally, if the wire is attached to the positive terminal of a battery, these electrons in the wire will tend to be

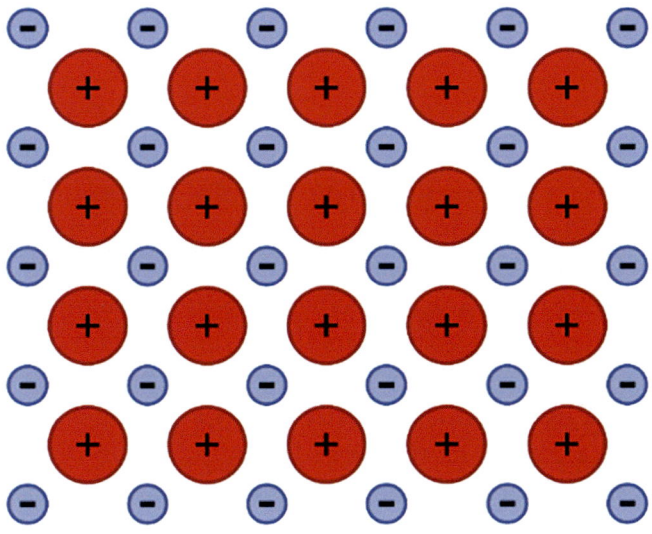

Figure 18.3 Metals: delocalised electrons (–) in a sea of ions (+). (Image credit: JackFromReedsburg via Wikimedia, Public Domain.)

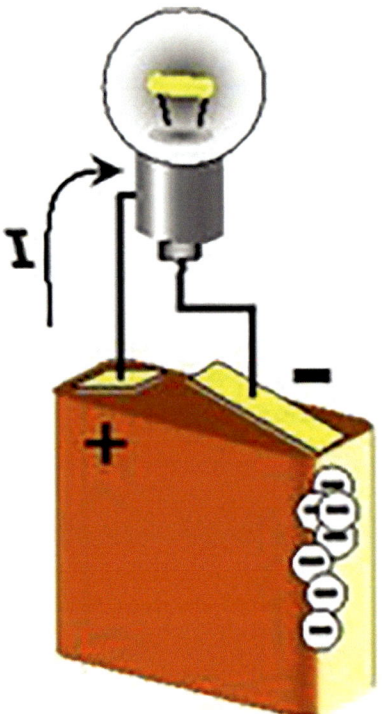

Figure 18.4 Current (I) flowing in a simple circuit.
(Image credit: Bela via wikimedia https://commons.wikimedia.
org/wiki/File:Circuit1new.jpg)

attracted towards the positive terminal. If a complete circuit of metal is now constructed with, say, a light bulb or heating element connected by wires to each end of the battery, all of these delocalised electrons will move from the negative towards the positive terminal of the battery, through the circuit. This constitutes an electric current (Figure 18.4).

Interestingly, the electric field is imposed throughout the entire wire more or less instantaneously at every point. Electrons everywhere in a wire will start to move simultaneously. That's why your lights switch on instantly – they don't have to wait for electrons to reach them! A good thing too, as electrons actually move extremely slowly along the length of a wire (typically, less than a millimetre per second), because they are constantly bumping into the fixed atoms of the metal – the eggs in the egg box – and losing momentum. It's mainly the electrons already inside your lights and machines that get propelled when you connect them up, not just the few arriving from a battery or through the mains.

Circuits can be broken by simply inserting a gap at some point – that's what switches do: breaking or completing a circuit, when they are off or on. In everyday situations, it is only in metals or a few other conducting materials that current can flow. In other materials, such as stone, air, glass, rubber, ceramics and most plastics, none of the electrons are free to move, so these act as insulators. For this reason, they can be used to make protective barriers in, for example, the plastic covering around a wire or the glass and ceramic insulators separating high voltage cables from the pylons that support them. When electric fields are extremely intense however, even air may break down and be able to conduct electricity – as lightning and sparks demonstrate.

PRODUCING ELECTRICITY

Looking back to our original quest, we were trying to work out where electricity comes from and how it reaches us. We've established that it will flow when freely moving, charged particles are placed in the electric field produced by a battery. The same holds when it is a remote power station, rather than a battery, that causes the current to flow. We have to ask now, how does switching on a light bulb or a mobile phone cause this to happen? The puzzling thing for one discussion group was how a washing machine or phone charger in your home links up to a power source, such as a power station, wind farm or solar farm, hundreds of miles away. It seems extraordinary that, by simply flicking a switch in your kitchen, two pieces of metal connect up and complete a circuit all the way to a distant generator. It's true: a single, effectively continuous line of metal stretches all the way from your household socket, under the road, through the substation, across the fields on pylons to the source. Your washing machine and a generator hundreds of miles away are linked in a single continuous circuit – and so are the machines of millions of others at the same time.

The way in which electricity is generated in a power station (or wind turbine or other kind) is quite different from that of a battery. It's not a chemical effect caused by layering a salty liquid between two types of metal. Instead, it's the result of an interaction between a metal

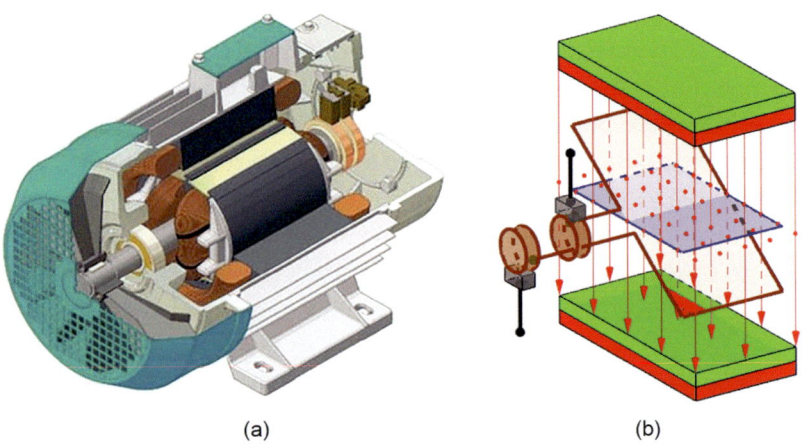

(a)   (b)

Figure 18.5 Generating electricity. (a) A generator.
(Image credit: Agonprebreza1 (https://commons.wikimedia.
org/w/index.php?title=User:Agonprebreza1&action=edit&red
link=1) via Wikimedia, https://creativecommons.org/licenses/
by-sa/3.0/deed.en).
(b) Copper wire rotating in a magnetic field. (Image credit:
MikeRun (https://commons.wikimedia.org/w/index.php?ti
tle=User:MikeRun&action=edit&redlink=1) via Wikimedia,
https://creativecommons.org/licenses/by-sa/4.0/deed.en)

and a magnetic field, moving relative to one another. Inside a power
station or wind (or other) turbine is a large machine called a generator
(Figure 18.5a) which produces electricity by spinning magnets around
in the presence of coiled-up copper wires (or vice versa – Figure 18.5b).
This causes an electric field to develop throughout the copper wire.
An animated version of Figure 18.5b is available at https://commons.
wikimedia.org/wiki/File:Electric-generator-animation.gif

When connected to external cables via the National Grid, this creates an
electric field throughout the entire connected system, from the overhead
cables, substations and transformers to the sockets and appliances in your
house. This electric field causes free electrons to flow throughout the system,
each energised by the electric field – what we know as an electric current.

ELECTRICAL ENERGY

Having established that the electricity that comes to our homes is a flow
of electrons in the various cables, wires and circuits of our electrical
equipment, it's still not clear what this has got to do with operating our

machines and delivering energy. 'What do the electrons do?' as Patrick put it in a discussion one day.

The important point is that electrons don't get 'used up' in an electrical circuit. They are simply pushed along the wires thanks to the force provided by a battery or generator. By this process, the electrons acquire energy – energy that was generated inside the battery or generator and transferred to the electrons. As the electrons pass through our various appliances, their energy is used to activate the machinery. For example: in a washing machine or vacuum cleaner, the energy of the electrons is used to turn the motor inside our equipment. In an electric heater or kettle, the energy is used to heat up the element. In summary, an electric current is a flow of charged particles each of which carries an amount of energy with it. Rather like trucks in a freight train conveying coal or sand, the charged particles travel along, picking up and offloading the energy they carry, but remaining intact themselves.

VOLTS, AMPS AND WATTS

This picture of household electricity as a flow of electrons through metal conductors, conveying energy to our various devices, helps us visualise what is otherwise an invisible process. With this concept in place, one discussion group turned, at this point, to more practical considerations: the words commonly associated with electrical equipment. 'What's meant by the Volts and Amps and Watts we read about on packaging and instruction manuals – what do we need to know about them?' asked Mary in one session. It's an important question because these terms appear in many places – on electrical goods and household bills, for example, yet are often misused in everyday communications.

What matters most to us, as consumers, is the amount of energy used in a given period – this is what we pay for (and it's what we raid the planet to source). It can also be important to know, when we plan to buy an item like a vacuum cleaner or light bulb, how powerful it is, whether will it work abroad and what fuse rating it requires. So let's start with these two: power and energy.

The *power* of a device or a light bulb simply means the amount of energy it uses every second – it's measured in Watts. A 100 Watt bulb is more powerful than a 50 Watt bulb – it uses up twice as much energy every second. A 1200 Watt vacuum cleaner will suck harder than a 800 Watt one. The amount of energy consumed (and therefore the cost on your bill) will, of course, depend on how long the bulb or cleaner is operating for. A 100 W bulb switched on for one hour will consume 100 kilowatt-hours of energy – that's what a kilowatt-hour (kWh) means: it's a measure of energy used. A 1200 Watt vacuum cleaner used for 10 minutes (a sixth of an hour) will use 200 kilowatt-hours. Electricity bills set out how many kilowatt-hours of energy you have used during the billing period.

'So why do we need to know about Volts and Amps then?' asked Julie, keen to minimise the number of terms she had to remember. A practical answer is that you may want to replace a fuse one day and need to check the rating: 3, 5 or 13 Amps; or wish to take your hairdryer abroad and check that it will work at the local voltage. A more enlightened answer is that it's always good to understand more of the world around you!

Everyday experience tells us a lot about flowing things: water circulating through radiators or filling up a bath, oil pushing through a car engine, rivers descending from the hills, for example. In each case, there are two interconnected factors to consider: the amount flowing past each second and the pressure it's under. A household tap is more pressurised the further it is below a water tank and as a result will flow more rapidly. Similarly, a stream may rush rapidly down a steep mountain slope, but meander at more leisurely a pace through lowland meadows. In the case of electricity, voltage is analogous to pressure and current to the flow rate. The respective units are Volts (V) and Amps (A).

To run the machinery needed in our lives, the key issue is its power rating – a few Watts (W) for a mobile phone, 1,000 W for an iron. Power means the amount of energy consumed each second, and it's the flowing electrons that convey the energy to your appliances. A given level of power can be achieved either by delivering lots of low energy electrons every second or fewer high energy ones. High voltage sources give electrons high energy, low voltages, low energy. Thus, a given amount of energy can be supplied at a higher voltage with a lower flow rate (Amps) or a lower voltages with a higher flow rate. The mains electrical energy for our homes is delivered at a fixed voltage – 230 Volts in the UK and European Union; 120 in the USA; 220 in China. So, to supply the required power for a kettle or oven in the USA, the flow rate or current must be almost twice as high as in Europe or China, the voltage being roughly half.

Talk of kettles and ovens inspired Ruth, in a discussion one evening, to ask why the plugs for our appliances today have three pins when once two pins were common. The answer is: to improve safety. Our equipment operates at relatively high voltages – 230 Volts in Europe. If a live wire inside an appliance were to come adrift accidentally and touch the outer casing of, say a kettle or oven, the latter would instantly be elevated from zero to 230 Volts. Were you or I to then touch it, a large current would be likely to flow through us to the ground, causing terrible, potentially lethal damage. The third wire in the plug connects directly to such exposed parts of an appliance and the corresponding third hole in the socket is connected directly to a piece of metal that penetrates the ground or earth outside. This 3rd wire is called 'earth' (or 'ground' in some countries) for this reason. It provides a path of much lower resistance than our bodies, so in the unlikely event of an electrical item becoming live on the outside, and us touching it, current would flow through this wire rather than us.

DISTRIBUTING ELECTRICAL ENERGY

Voltage also matters when it comes to getting electrical energy out to where it's needed. All conductors of electricity, even metal wires, resist the flow of current to some extent. The flow of electrons is impeded by collisions with atoms in the metal (or more strictly, ions – atoms lacking one or more electrons). It's this opposition to the flow of current that is what we call electrical resistance. The cables delivering electrical energy from power station to users are no exception to this. Good conductors though they are, the resistance they offer to the flow of current is nevertheless significant because they are so long, stretching sometimes over hundreds of miles. Where there is resistance, some electrical energy is always lost as heat. This occurs because the fast moving electrons transfer part of their energy to the atoms (strictly, the ions) of the metal as they collide with them. Fortunately, heat losses such as these are reduced when the energy is conveyed at a higher voltage.

To send electrical energy across the country, the voltage is stepped up between the generator and the long transmission lines. This reduces losses through heating due to the resistance of the cables. It then has to be stepped down again locally, ready for safe use in homes and factories. The large transformers that step voltages up and down can be seen at the electricity substations dotted around our towns and rural places (Figure 18.6).

Today, transformers are often used inside our homes to step down voltages even further. The devices we use to charge our mobile phones (cell phones) and computers reduce the voltage of the electrical energy received through the 230V mains to the few Volts that's needed for our digital equipment.

Figure 18.6 Electricity substation.
(Image credit: Danny P Robinson.)

ELECTRICITY AND ELECTRONICS

Mention of mobile phones and computers prompted a basic question in one discussion group – a question often pondered but rarely asked. 'What's the difference between electricity and electronics?' Sarah wondered. 'Is it something to do with the voltage?' she added. It is indeed: electronic devices need very little energy each second (power) and a low voltage suffices to deliver this. By contrast, electrical devices like washing machines and room heaters need much more power in order to turn motors round or produce the heat needed to warm a room. The power rating of a mobile phone, for example, may be a few Watts, whereas that of a washing machine or iron is closer to a thousand.

The role of an electronic device is to control relatively small flows of electrons in tiny electrical circuits. Controlling such flows, in very rapid, orderly succession, enables logical operations like arithmetical calculations and changes to pixels on a screen, to be regulated. This kind of control is the basis of computers, phones, digital screens and the host of electronic devices with which we organise our modern lives. In the early days of electronics, the circuits needed for this kind of control comprised relatively bulky components, including capacitors, resistors and valves. You can see examples of these today in museum exhibits of early televisions and computers. Since the discovery of semi-conducting materials, like silicon, and the subsequent invention of the transistor shortly after the Second World War, rapid progress has been made in

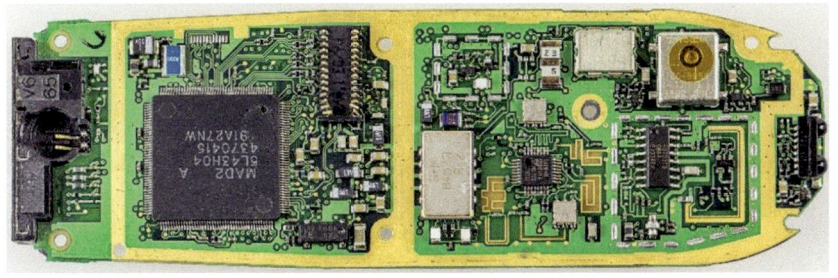

Figure 18.7 Circuit board in a mobile (or cell) phone.
(Image credit: © Raimond Spekking / CC BY-SA 4.0
(via Wikimedia Commons) https://creativecommons.
org/licenses/by-sa/4.0/deed)

miniaturising circuits by etching them in specially prepared boards (see Figure 18.7). These are the wonderful and complicated-looking boards to be found in much electronic equipment today.

CONCLUSION

This chapter has tackled many quite tricky fundamental concepts about an extraordinary invisible presence in our lives: electricity. It's not easy to grasp the meaning of words such as current, voltage, energy, power and resistance. It may take more than one reading to get the hang of these ideas. Once grasped, however, many aspects of everyday life become less baffling.

Extensive though this chapter has been, many other important questions have still not been addressed. What, for example, are AC and DC (Alternating and Direct Current)? Why are electric shocks dangerous? Are nerves electrical? How does a motor work? What are microchips? How do solar panels work? Most important of all, perhaps, how is electrical energy going to be generated in future as the world moves away from burning fossil fuels? How is it going to be distributed to the places it's needed? How can electrical energy be stored for powering cars, aeroplanes and ships? It's clear the story of electricity is rich and complex. Its origins lie with the inquisitive explorers and ingenious inventors of two and three centuries ago; its future will determine the health or otherwise of our planet. Hopefully, the fundamental concepts introduced in this chapter will prove useful as you try to make sense of whatever developments are to come.

Nineteen

THIS BOOK

This book has introduced concepts from a range of scientific disciplines, based on questions and observations from everyday experience. The purpose has been to make important ideas in science accessible, enhance understanding and help build confidence when engaging with conversation and media stories involving the sciences. It focusses on issues people have raised in discussions, rather than working to build up knowledge systematically from elementary beginnings. As a result, advanced concepts have sometimes been introduced without first explaining simpler, underlying ones. For example, DNA is described in relation to heredity without the components that make up the DNA molecule having been introduced first. A further consequence is that the sequence of chapters appears somewhat haphazard. This reflects the way discussion often evolves over the longer term: jumping from topic to topic in response to everyday matters, rather than digging, ever deeper, into one particular issue.

ENGAGING WITH SCIENCE AS AN ADULT

These aspects of the approach taken in this book contrast with the carefully structured syllabuses followed by students in school and university. The latter are, quite properly, designed to ensure that knowledge acquired in earlier stages can be built upon to explain more complex ideas later on. For many adults, however, a different approach is needed. Wishing simply to exercise their curiosity about the world, they may not have the time or inclination to build up foundational knowledge in this gradual manner. They may, for example, wish to find out more immediately about the nature of energy, without the full back-story, because (a) they are concerned about the environmental impact of producing it and (b) they are paying a lot for it every quarter.

A consequence of this more spontaneous approach is that the picture of science that emerges over time can be somewhat kaleidoscopic:

DOI: 10.1201/9781003272779-20

colourful but fragmented. Knowledge acquired in this way may initially appear patchy and introductory, but, over time, it can develop into something more coherent as key concepts begin to reappear repeatedly in varying contexts. The structure of molecules, for example, crops up in such apparently unrelated topics as glues, heredity and vaccination, even though it was not the starting point for any of them. For the curious layperson, creating a storehouse of factual knowledge is not necessarily the principal aim. Their reason for engaging with science may be quite different from that of a student working towards a science-based career, for whom a well-stocked and orderly compendium of facts and concepts is essential. For people who gave up on the subject early in life, becoming more confident about scientific ideas may be the main benefit. By rediscovering an interest, they may find themselves grasping concepts that had eluded them at school. Renewed confidence can help people follow news stories more thoroughly, engage in, rather than avoid, conversations about science and challenge any bluff and bluster they may perceive.

It's not only confidence that may be lost through a patchy experience of science education; nor is it just pieces of knowledge: quiz-show items, such as the elements of the Periodic Table or Latinised names of bones. A deeper loss can be of some kind of schema or conceptual landscape into which new ideas can be fitted. Thus, for example, the apparently unrelated experiences of sunlight, radio waves and X-rays may not be perceived in terms of the unifying concept of electromagnetic waves. Useful connections between discrete areas of knowledge may also be missed, in the absence of an overall schema: for example, between the concept of electromagnetic waves and the functioning of the retina.

Long experience of discussing science with groups of adults has informed the approach taken in this book. By starting from a matter of immediate concern, people tend to be more motivated to both engage with an issue and pursue it until it's understood. By following the path of questioning as it occurs in discussion, rather than the usual logic of a text book, interest is kept alive. When curiosity drives learning, discussion regularly moves across the academic disciplines, as it evolves: from the microbiology of viruses and anatomy of the respiratory tract to the mathematics of epidemics and physics of air currents, for example.

Over time, connections begin to appear in what can, initially, be a somewhat fragmented picture of scientific knowledge. Whether you start from the action of antibodies or the air pressure in a cabin, you soon get down to molecules and how they interact with one another. Heating your home or digesting your food, both bring you into contact with the concept of energy. These are powerful concepts, capable of illuminating huge swathes of your daily experience. Regular and repeated acquaintance with such ideas, albeit in utterly varied contexts, helps reinforce learning. Grappling with atoms, cells or forces from dozens of different starting points enables a rich and memorable impression of their meaning to develop.

It's not only a sense of the landscape of scientific ideas that can build up in this way but also a more sophisticated concept of how science itself works. By exploiting the interplay between scientific knowledge and personal experience, connections may be made with the economic, social, political, philosophical and religious ideas that adults acquire as they mature. A memorable discussion in one group began with how medicines interact with the body. Introduction of the word 'drug' in a medical context soon led to the issue of illegal drugs and ultimately to a session with the chief scientist at the Home Office (Interior Ministry) about the relationship between science and public policy. Other discussions that began with apparently technical questions about mass or electric charge have led into more profound consideration of philosophical or religious perspectives on the ultimate nature of reality.

ABOUT SCIENCE ITSELF

Re-engaging with science as an adult brings an important extra reward, beyond new concepts and insights into the world around us: it also opens up questions about the way science itself works. The interaction of scientific knowledge with economic, political and social forces has been brought into sharp focus by the pandemic. Political decisions about which lines of research to fund and societal choices about adherence to public health measures, such as vaccination and mask-wearing, have drawn upon and, in turn, shaped scientific enquiry.

Science and the technology it spawns are not simply switched on at short notice because an immediate problem needs to be solved; they develop cumulatively over long periods of time. Understanding of antibodies, T cells and other elements of the immune system has been building up continuously over decades, long before the pandemic shone a spotlight on the field. Advances in gene manipulation and nanotechnology meant that delivery mechanisms for viral antigens were ready to go, as soon as a new type of vaccine needed to be developed. Once the political and economic decision had been made, it was no accident that it proved possible to produce vaccines at breakneck speed. The pooling of accumulated knowledge and expertise between virologists, molecular biologists, chemists, engineers and pharmacologists around the globe was critical to solving the practical problems. Team-working and collaboration are essential aspects of science in action.

Finding solutions to problems through science is not, however, only a matter of persuading authorities to invest. Making progress in scientific research entails multiple lines of enquiry and fearless rejection of those that fail or draw a blank. Scientists have to develop a sceptical attitude to their own conjectures. As one philosophy of science sees it: the task of a scientist is to try to *disprove* their pet theory, rather than just back it up. Experiments are designed to test a proposition by finding as many ways as possible to show it to be false. Only when an idea survives these tests is it taken seriously by other scientists. Even then, it takes a community of specialists – the peers – to refine and finally approve scientific findings as valid. Scientists are as subject to human frailties as the rest of us: they may feel as rivalrous, vain or ambitious as any of us, but at least the discipline within their culture militates against excess in these directions.

Public exposure to real science has also illuminated something of the methods of science. Statistical analysis of drug trials and infection rates informs public policy, biochemical experiments lead to potential vaccines, air flow measurements inform ventilation systems, DNA analysis pins down differences in emerging variants. These widely varying methods are just a sample from the many disciplines contributing to one urgent public health problem. Science, as a whole, stretches to methods way beyond those of the laboratory and clinical setting: from field trials in agriculture to remote imaging in astrophysics. This extraordinary diversity of methods and purposes is characteristic of science as a whole. Each field has become increasingly specialised and scientists correspondingly so. What is common across the specialisms is a sense of exploration: the uncertain, the unknown, the possible, the unlikely; all are central to the practice of science.

WHERE DO WE GO FROM HERE?

Where does all this leave the layperson with a general interest in science – such as they might have in politics or the arts – but not seeking a career or even hobby in it? Is there no role for them today, in this world of highly specialised knowledge and expertise? If you missed the boat at school, is it simply too late to develop an interest in the subject? My experience of working with people in this position leaves me with a few conclusions.

Firstly, it's clear to me that it's both possible and beneficial for adults to engage with science, in one way or another, regardless of previous experience. For some, this means pursuing a particular interest in, say, genetics or artificial intelligence through specialised books and courses. Such resources are now widely available in accessible formats. For others, a broader approach is preferred that is related more closely to their experiences and observations in life. My experience with groups of such people meeting in informal circumstances is that discussion, based on real-life issues, though anecdotal initially, can indeed lead into serious discussion about scientific concepts. Direct interest in the issues raised can strengthen motivation to learn and deepen engagement with the concepts.

For an adult wishing to try to understand more of science, the key seems to me to be clear and realistic about what one is looking for. Certainly, a career in scientific, technological or medical fields does require accumulation of basic knowledge over many years of study, followed by further years of specialised training. But that is not usually what the layperson, driven simply by curiosity, wants. For the ordinarily curious, for whom this book is intended, the benefit of engaging with science is simpler and humbler. It can bring the joy of seeing how something works or something can be explained. Hard on the heels of this comes a growing confidence and sense of achievement in grasping scientific ideas previously thought to be beyond reach. Knowledge gained in this way, later in life, may appear patchy and somewhat arbitrary, but, couldn't much the same be said for that which we acquire and retain into adulthood from formal schooling?

Cultivating an interest in science as an adult does involve overcoming a number of obstacles, of course. Some books, even for the layperson, use overly technical language or are too specialised; a TV documentary may seem over-simplified or condescending, a radio programme too jazzed-up. Choices have to be made about what to read, watch or listen to and the same goes for courses, exhibitions and festivals. You can only try out a variety of resources, find your preferences and discard those that don't work for you. The essential point is not to give up at the first hurdle. Whatever it is that defeats you, there'll be an alternative that suits you better. Some suggestions about available resources are given in the last few chapters of my earlier book 'Getting to Grips with Science: a fresh approach for the curious'.

CONCLUSION

Some kind of appreciation of science seems to me to be increasingly important, as the consequences of climate change begin to impinge on our lives. As adults, we'll be voting for policies, working in communities, influencing our businesses and adapting our lifestyles in efforts to reduce emissions of greenhouse gases. It's a great pity that formal science education (from the upper secondary level) has offered so little, historically, to those with no career interest in the subject. Despite this, there remain many ways of re-engaging with science, in a different way, as an adult. The subject is on the rise in the media, with regular TV documentaries on medicine, geology, evolution, astronomy, engineering and the like, and discussions, analyses and even science-comedy mash-ups flourishing on the radio. Popular Science shelves in bookshops are well stocked with new titles appearing regularly. Scientists themselves are reaching out as never before, through exhibitions, visits, festivals and pub-based discussions.

The effort being made, by so many people, to follow scientific developments, grapple with unfamiliar concepts and make good their understanding of science is inspiring. If scientists and citizens continue to reach out to each other, the world will surely be a better and safer place. If you've managed to read this far, you are already making an important contribution to this cause.